Community Based Water Management and Social Capital

Community Based Water Management and Social Capital

Kiyoshi Kobayashi, Ibnu Syabri,
Ismu Rini Dwi Ari and Hayeong Jeong

IWA PUBLISHING

Published by **IWA Publishing**
Alliance House
12 Caxton Street
London SW1H 0QS, UK
Telephone: +44 (0)20 7654 5500
Fax: +44 (0)20 7654 5555
Email: publications@iwap.co.uk
Web: www.iwapublishing.com

First published 2014

British Library Cataloguing in Publication Data
A CIP catalogue record for this book is available from the British Library

ISBN: 9781780405452 (Paperback)
ISBN: 9781780405469 (eBook)

Contents

Chapter 4

Chapter 5

List of Figures

List of Tables

About the Authors

Kiyoshi Kobayashi, BEng., MEng., Dr.Eng., FJSCE, Professor of Graduate School of Engineering, Kyoto University, Japan and Professor of Graduate School of Management, Kyoto University, Japan. He serves on the editorial boards of the American Society of Civil Engineers, the Annals of Regional Science, the Journal of Applied Regional Science and the Japan Journal of Civil Engineering. He is also the editor in chief for the Journal of Japan Society of Civil Engineers and Vice Editor in chief, Journal of Infrastructure Systems. He is the author and co-editor of 38 books and more than 380 technical papers.

Ibnu Syabri, BSc., MSc., PhD., Associate Professor at School of Architecture, Planning and Policy Development, Bandung Institute of Technology. His research interests are Regional Science, Remote Sensing, GIS, Transport Simulation & Modeling. His works have been published in several international journals and conferences including the Journal of Geographical Analysis.

Ismu Rini Dwi Ari, BSc., MSc., Ph.D, Assistant Professor at Regional and Urban Planning Department, Engineering Faculty, University of Brawijaya, Indonesia. Her current research interests are related to water governance, rural and urban poverty, community development planning and social capital. Her works have been published in the international conference proceedings, including the Procedia Environmental Sciences, Elsevier.

Hayeong Jeong, B. Eng., M. Eng., Dr. Eng., MJSCE, MIEEE, Program-Specific Assistant Professor, Chair of Human Resource Development for Asian Business, Graduate School of Management, Kyoto University, Japan. Her current research interests include public involvement, NGOs, corpus data mining, rural development and social capital. Her works has been published in the international journals and conferences including the Journal of Computers and Journal of Japanese Society of Civil Engineers.

Gautam Ray, M.Sc. LL.B. MA(Econ) and Ph D., Professor of Graduate School of Management, Kyoto University, Japan. His current research interests include legitimacy theory, collective action theory and infrastructure development

Kakuya Matsushima, Dr. Eng, Associate professor at the department of Urban Management, Graduate School of Engineering, Kyoto University. His major is infrastructure economics, transportation economics, and social capital. He was awarded Japan Society of Civil Engineers young researcher award in 2002, and Sakashita Prize from Applied Regional Science Conference in 2009.

Kenshiro Ogi, He received BEng. and MEng. in urban management from Kyoto University. He joined Rakuten Inc in 2012. Currently, he has been working in the Group Marketing Department and is engaged in planning the whole Rakuten group marketing strategy.

Hans Westlund, PhD in Economic History (Umeå University 1992), Professor of Regional Planning, Royal Institute of Technology, Sweden, Professor of Entrepreneurship, Jönköping

International Business School, Sweden, and Professor at the Institute for Developmental and Strategic Analysis, Ljubljana, Slovenia. He is President of the Western Regional Science Association 2013/14 and Chairman of the Epainos Committee of the European Regional Science Association, with the task of rewarding best papers of young regional scientists.

Tsuyoshi Hatori, Dr. Eng., Associate Professor of Graduate School of Science and Engineering, Ehime University, Japan. His current research interests are related to public debate and consensus building, social capital, and regional governance. His work has been published in several international journals and conferences related to Civil Engineering, Transportation and Social Psychology.

Raymond Valiant Ruritan, BSc., MSc., Professional engineer, certified under the Indonesian Association of Hydraulic Engineers and the Indonesian Committee on Large Dams. His current position is Director for Technical Affairs at Jasa Tirta I Public Corporation, a state-owned company that serve as a river basin organization for the Brantas and Bengawan Solo River Basins in Java, Indonesia. His main interests comprises river and fluvial engineering, economic engineering, hydro-social sciences and development history.

Masamitsu Onishi, Dr. Eng., Assisitant Professor at Graduate School of Engineering, Kyoto University, Japan. His research interests include institutional issues related to infrastructure management from the institutional economic perspective. He was awarded Japan Society of Civil Engineers young researcher award in 2002. He has been published more than 40 referred papers in the academic journals including Journal of Japan Society of Civil Engineers.

Mangapul L. Nababan, BSc., MSc., Dr.Eng., He works at Center For Strategic Studies, Ministry of Public Works of the Republic of Indonesia. His works include studies on urban and regional infrastructure management and infrastructure strategic planning. His current research interest includes decentralization, local public goods provision and water resource economics.

Adipandang Yudono, BSc., MURP, Lecturer and Researcher at Regional and Urban Planning Department, Engineering Faculty, University of Brawijaya, Indonesia. His research interest areas are Watershed Management, Disaster Management, Coastal Management and Mass Rapid Transport System Modelling. He got research funding from USAID for Waterspring Conservation Management in Malang Regency, Indonesia (2009), and Indonesia's Ministry of National Education for Tsunami Disaster Management in Nusa Tenggara Timur (NTT), Indonesia (2010) and for Tsunami Disaster Management in Pacitan, Indonesia (2012).

Arief Rachmansyah, B.Eng, Dr. Rer nat, Senior Lecturer at Department of Civil Engineering, Brawijaya University, Indonesia. His main research interests are geology and disaster management. Currently he is the chairman Indonesia Geologist Association for East Java region.

Sri Sudaryanti, BSc., MSc., Senior Lecturer at Department of Water Resource Management, Faculty of Fisheries and Ocean Science, Brawijaya University, Indonesia. His main research interest is social capital for water resource management.

Sudarto, BSc., MSc., PhD, Senior Lecturer at Department of Agrotechnology, Faculty of Agriculture, Brawijaya University, Indonesia. His main research interest are land capability and suitability for agriculture purpose, GIS and Remote Sensing modelling for agriculture.

Surjono, BSc., MTP., PhD., Associate Professor at Regional and Urban Planning Department, University of Brawijaya, Indonesia. His research interests are rural and urban development. His works have been published in the international conference proceedings including the International Union for the Scientific Study of Population. He is a member of Indonesian Planner Association and active in sharing knowledge in national and international conference, writing text books and articles in academic journals related to urban and settlement planning.

Mustika Anggraeni, BSc., MSc, lecturer and researcher at Regional and Urban Planning Department, University of Brawijaya, Indonesia. Her research interests are related to climate change and its impact to urban settlement. She participated in climate change international workshops at Rotterdam (2012) and Trieste (2013).

Joni Hermana, BSc., MSc., PhD., Professor at Department of Environmental Engineering, Faculty of Civil Engineering and Planning, Institute of Technology Sepuluh Nopember, Surabaya. His main research interest is waste water management. His works have been published in several international journals and conferences including the International Biodeterioration & Biodegradation, Elsevier.

Preface

The subject of this book is community-based water management in developing Asian countries, underlying social capital, and participatory approaches to water management initiated by rural and suburban residents in order to overcome the problem of the exclusion of vulnerable and small users from public water service.

By introducing the latest field experiments and theoretical studies in economics, social science, engineering, and regional planning, including game theory, microeconomics, econometrics, statistics, social network analysis, social choice, and micro-finance, this book provides a scientific understanding of community-based water management—how to secure responsible management to satisfy quality and quantity requirements, as well as how to synchronize community-based water management with public water service.

This book is relevant to many—to citizens and planners, to the 6 billion people in Asian countries who suffer from water scarcity, and to all the people in developing countries in Africa, the Pacific Islands, Central and South America and Central and Eastern Europe who are in short supply of public water service.

The stimulus to publish this book came from the IWA during the international conference series on Water Supply Management and Social Capital, held by Kyoto University in collaboration with Brawijaya University, Bandung Institute of Technology, Sepuluh Nopember Institute of Technology and Hasanuddin University, from 2009 to 2012. Hosted by four Indonesian universities, the conference began at Bandung city in West Java Province, continued at Surabaya and Malang cities in East Java Province, and closed at Makassar city, Sulawesi Island. The conference facilitated the acknowledgement of ideas and built a communicative framework by which researchers and practitioners can contribute ideas or help solve problems related to such issues as (i) Social Structure and Institutions in the Community, (ii) Water Usage Rights and Water Governance, (iii) Water Policy and Organization Agencies, (iv) Community-Managed Water Systems, (v) Sustainable Water Management and (vi) Future Perspectives.

The chapters of this book are revised outstanding papers selected from the conferences. The fourteen chapters in this book introduce novel concepts and theories of community-based water management. Additionally, this book draws upon various empirical cases reflecting Indonesia's water problems, including coordination among municipalities, community-managed water supply systems in rural areas, community participation in conflict resolution, and conservation activities related to water resources.

Finally, deepest appreciation must be expressed for the international publication of this book, which will serve as a reference for researchers as well as for practitioners throughout the world who deal with water issues and problems. Moreover, this book will enrich the collection of textbooks related to community development planning and social capital, particularly with respect to cases in developing countries.

Kiyoshi Kobayashi
Ibnu Syabri
Ismu Rini Dwi Ari
Hayeong Jeong

Foreword

The publication of this book comes at a timely moment, when water supply problems are affecting many communities around the world. This book makes an important contribution by providing information and examples on how communities collectively can address their lack of access to water services. The book presents valuable examples from empirical cases on how people have organized themselves to overcome problems of exclusion, on how a common voice can move a community's voluntary association into action to increase water services, and on how cooperation between individuals can achieve sustainable water management based on their capacity weaknesses.

Appropriate water governance is one of the main issues that should be addressed to deal with the world's water problems. This book provides significant evidence on effective solutions to water crises, based on case studies and real-world practical experience. In addition, the book is equipped with a theoretical perspective. In the end, this book will be an important addition to collections of reference books promoting the understanding of human cooperation in water management.

Moch Bisri
Professor in Water Resources Engineering Department
Dean of Engineering Faculty
University of Brawijaya
Indonesia

1 Aspects of community-based water management and social capital

Ibnu Syabri[1], Kiyoshi Kobayashi[2], Hayeong Jeong[3], Ismu Rini Dwi Ari[4]

[1] School of Architecture, Planning and Policy Development, Institute of Technology Bandung, Ganesha 10, Bandung, Indonesia 40132. E-mail: syabri@pl.itb.ac.id
[2] Graduate School of Engineering, Kyoto University, Katsura, Nishikyo-ku, Kyoto, 615-8540, Japan. E-mail: kkoba@psa.mbox.media.kyoto-u.ac.jp
[3] Graduate School of Engineering, Kyoto University, Katsura, Nishikyo-ku, Kyoto, 615-8540, Japan. E-mail: hayeong@psa2.mbox.media.kyoto-u.ac.jp
[4] Engineering Faculty, University of Brawijaya, Malang, Indonesia. E-mail: dwiari@ub.ac.id

Abstract

The purpose of this introductory chapter is to examine what is meant by community-based water management and social capital. This leads to a discussion of the various ways in which the study of community development and social capital can be approached. The contributions in this volume reflect the development of community-based water management (CBWM) in a number of ways. These include the voluminous literature devoted to problems of CBWM regarding water scarcity and accessibility. Increasingly, specifically in Indonesia, the attention have been directed to the implementation (i.e., policies and strategies) of CBWM, such as how individuals or communities establish a collective action whereby they work together to improve water services, a roadmap or valuable guidance on how groups can become organized to influence decisions that affect their connectivity and relationship to a common goal. Finally, this volume provide useful background as well as a stimulus to pursue future research on community based-water management and social capital.

Keywords: collective action, social capital, community-based water management

1.1 INTRODUCTION

UNICEF (2007) in Global Environmental Outlook reports that about 655 million people in developing countries, i.e., 17.6 percent of the world, still lack access to safe water; about 1.8 million children die annually; and about 433 million school days are missed due to diarrhea from unsafe water and poor sanitation. In the case of Indonesia, among a total of 231.6 million inhabitants, only around 49.7 percent of citizens have access to water, 10 percent of which is obtained through community-based water management (BPS, 2005). These figures are evidence that the problems of a rapidly growing world population and increasing water scarcity have become of considerable magnitude. Westlund (2006) argues that it is important to recall that water scarcity is only partly a problem of physical lack of water, and to a much larger extent water shortage is a poverty problem, a lack of government capacity, and a lack of economic resources to build the necessary structures for water supply. Nevertheless, managing water resources at a global or state level can be over-ambitious and unrealistic, particularly when many developing countries have weak regulating institutions and limited technical and financial capacity.

There is a need to change the mechanism for water resource management to give greater respect to needs and priorities, particularly to users on local geographical scales, involving communities which are being threatened by a lack of access to water. Therefore, it is felt that user communities must have a greater share in determining the demand and supply of water through an inclusive and participatory approach. In this process grassroots communities will become a part of the water management planning and implementation process. Community-based water management (CBWM) has gained worldwide acceptability and recognition over the years, and local knowledge on traditional water management practices has become the center stone of this approach (Devine, 2006).

The purpose of this introductory chapter is to examine what is meant by community-based water management and social capital. This leads to a discussion of the various ways in which the study of community development and social capital can be approached. The contributions in this volume reflect the development of CBWM in a number of ways; in addition, there has been a significant deepening of the understanding of the way in which the collective action of stakeholders (e.g., water users and government institutions) can be developed. Consider the voluminous literature devoted to problems of CBWM regarding water scarcity and accessibility; increasingly, it has become clear that some of this attention might have been better directed to the implementation (i.e., policies and strategies) of CBWM, such as how individuals or communities establish a collective action whereby they work together to improve water services, a roadmap or valuable guidance on how groups can become organized to influence decisions that affect their connectivity and relationship to a common goal. Particularly, how is the concept of "synchronizing community-based water management with public water service" realized? Can a community just drill a well and start providing service, or might that water or the water rights "belong" to a larger utility? How or why might a community with their own small system integrate with larger water districts? What are the benefits and pitfalls associated with synchronizing, and how does it occur?

The present volume originates from the above considerations, and attempts to provide a prospective view of novel concepts, theories and empirical aspects of CBWM. It should not be regarded as simply a collection of ongoing research. Rather, it is a selection of ideas and reflection put forward by scientists and professionals exploring new insights in the field of community development planning.

The contributions of the book are organized into four main sections:

- The foundation of collaboration
- Water governance and participatory approaches
- Institution of community-based water management
- Case Study

1.2 FOUNDATION OF COLLECTIVE ACTION

The theoretical foundation of community-based water management is based on the collective action or collaboration concept, which hypothesizes that problems within communities can be overcome if people collectively develop endogenous behavioral choice based on human network and shared knowledge, i.e. social capital, about the beneficial effects of such action (Ray, 2012). Social capital can be defined as non-formalized networks that are used by the networks' nodes/actors to distribute norms, values, preferences and other social attributes and characteristics. In this way, social capital forms an informal counterpart to formal institutions (rules) and organizations. Another way to describe this is that social capital is a type of infrastructure (with nodes and links) and "traffic" (of information, values, etc.). This means that both the quality of the social links and values, and the quantity of them matter (Westlund, 2006).

According to Ray (2012) , the accumulation of social capital and its accessibility are critical determinants of the failures or successes of collective action. However, the accumulation and dissemination of social capital is costly for a number of reasons: i) market failure because of externalities, public good characteristics, and informational asymmetry; ii) powerful vested interests (which often include state machineries) preferring the status-quo and obstructing the acquisition/dissemination of such knowledge; iii) inadequate supply of focal actors with strong enough preference for their non-material well-being to bear the cost of acquisition/dissemination of such knowledge and organize a network of potential actors. Therefore, for sustaining a collective action, Westlund (2012) suggests that the society must be endowed with a group of people having a strong enough preference for enlightened self-interest to be willing to incur the cost of establishing a social capital through which such capital can be disseminated, or spilled over to other potential actors of collective action. Social capital is a multi-faceted concept and connected to all human interaction. Therefore social capital is important in all growth and development policies. No matter which water supply solution is implemented or how many users there are, good governance of the water supply system is necessary, and good governance arguably cannot be achieved without good social capital.

A careful combination of top-down and bottom-up methods is necessary, i.e., a unique combination for each case. Participatory planning is one way to achieve the good governance that produces good social capital, but there are very few experiences above the very local, neighborhood or village level. Civil society is often underdeveloped outside the western world. NGOs often represent ideological, economic or donors' interests. This means that they are not a guarantee for "good governance," even if they can play a very positive role in implementing concrete projects. Policies supporting citizens' own engagement, and against corruption, are best for fostering the good governance and social capital that promote efficient water management systems. Moreover, sustainable, efficient water management systems—self-organized, public or private—need networks for information exchange, negotiations and trustful management; i.e., they need social capital.

Self-organized systems for water management are most often rural phenomena. What aspects of social capital are important in such rural development processes? As for social capital and governance, the relationship between the two is mutually reinforcing. Good governance improves bridging and linking social capital. Good bridging and linking social capital improves good governance, i.e., openness and transparency, inclusiveness and communicativeness, coherent and integrated policy, equitability, accountability and efficiency.

The history of water supply and provision in Indonesia gives us a unique experience on how over time culture has influenced the community-based water management system.

Ruritan (2012) explains that water governance in Indonesia cannot be separated from the history of Indonesian culture that consists of diverse ethnic groups geographically spread across the archipelago. The history of water governance in Indonesia can be traced back from early Hindu-influenced kingdoms in the sixth century on the island of Java, to the Indo-china and Islamic and western colonial periods, and ultimately to the independence phase of the Indonesian state beginning in 1945. In explaining the dynamic relation between water and society, Ruritan argues that though Java is a very dominant tribe, this does not mean that ethnic Javanese culture dominates water governance in Indonesia. Nor does such dominance provide sufficient arguments in favor of a particular type of culture. At the pre-colonial period, Indonesia was characterized by less population density. Water was governed mainly at the village and free-estate levels under landlords and local authorities. In general, at that time, water infrastructure was funded by (mostly) the commons. Organizing construction activities and providing funds for operation and maintenance were tasks borne mostly by the users. A certain water fee was recorded in the year 1486, in what can be seen as a precursor of the application of a cost recovery scheme for water service. In the period of Western colonialism (1700s – 1940s), water governance evolved rapidly, especially after social ethics and technology development were introduced. In terms of water governance, rational planning and efficient administration are two important pillars left by the Dutch Indies colonial administration. Even after Indonesia became an independent state, these pillars remained as reminders of the Dutch legacy. After Indonesia became an independent state, the hierarchic approach in water governance was carried over by the newly established Indonesian Government. However, following the passing by the Indonesian parliament in 1999 of Laws 22/1999 and 25/1999 concerning regional autonomy and fiscal decentralisation, a new water law was introduced in 2004, Law No. 7/2004. The essence of the law is a shift of several government functions, responsibilities, and tasks, including water governance, from central to local governments, so that local governments, including provinces, regencies and municipalities would have a greater role in several important functions, including water supply provision. This implies that while on the one hand the government has to change the way it manages resources, on the other hand the people are requesting more control over the available resources. During the reform period, people also had to change themselves into a looser grid-relationship, moving towards more egalitarian ways. In this way, the power of the collectives, when they speak with a common voice, shall have more power to pressure management into action, such as increasing services or reducing fees, or even changing allocation. As guaranteed by Law No. 7/2004, water resources management crosses sectors and areas of interest, and the egalitarian way requires cohesiveness of action to maintain the sustainability of the function and benefit of water resources. Nowadays, as a platform for stakeholder liaison, cohesiveness of action and advisory mechanism for water-related basin-level development and management, a coordinative and consultative mechanism is required to integrate the interests of various sectors, territories and stakeholders in water resources.

1.3 WATER GOVERNANCE AND PARTICIPATORY APPROACHES

The implementation of community-based water management varies widely across regions in Indonesia. Styles of local governments (i.e., regencies and municipalities) range from fully centralized to fully decentralized, with many variations in between. There are regions where water resources are managed solely by the government, regions where they are managed by the private sector or communities surrounding the resources, and regions where they are managed jointly by the government, the private sector and the communities. For many areas in Indonesia, implementation of community-based water management policies is central to achieving sustainable water management. It would be useful to understand why some regions or communities have successfully developed these services whereas others have not.

Nababan et al. explore asymmetric ownership between municipalities in water supply system provision in Indonesia. This is an example of a problem in water governance policies between jurisdictions. Following the decentralization process in Indonesia, local governments' role has become increasingly important in terms of developing policies for regional public infrastructure. Water supply system provision has distinguished characteristics in terms of jurisdictions and characteristics of regions which affect the quality and quantity of water delivered to residents. Management of a state-owned enterprise in water supply system provision becomes complex, especially when the formation of new municipalities results in a need to separate assets in a fair way. In the case of PDAM Tirta Raharja (i.e., a state-owned enterprise), ownership belongs to Bandung Regency, whereas its coverage service includes Cimahi City and West Bandung Regency. Regarding water supply provision, the authority to determine production level and price is vested in the regent or mayor of the city. The Decentralization Law has stipulated that water exploitation authority is determined based on jurisdiction; however, due to characteristics of geographical areas, some regions still rely on other regions to supply raw water as input for clean water production. In this sense, local government has the authority to use the water resource, for example by exploiting the surface water, groundwater, and spring water in its jurisdiction, to supply clean water to its residents. For surface water such as from a river involving 2 or more regions, this could be considered as a provincial river under provincial government authority, but for other sources such as groundwater and spring water, the permit system is more complicated due to intersectoral issues, for example coordination between the forestry and mining offices at each respective government level. Hence, water use right regulation in terms of cross-jurisdiction water transfer is still debatable, and until now conflicts between jurisdictions have been resolved solely by bilateral or multilateral agreement. However, in the case of conflict between cities or regencies, provincial government can play a role as mediator and facilitator, as stipulated in the law.

In order to overcome the problems found in asymmetric ownership of water resources between government institutions, Onishi et al., in "Partial Authority Allocation of Regional Water Supply System in Indonesia and Economic Efficiency," report that partial authority allocation can be problematic when regional asymmetricity regarding the availability of alternative water sources and seasonal fluctuation exists. The price of the networked water is higher than the efficient level and investment in pipelines in the urban area is insufficient, whereas in rural areas it is excessive. Hence, the insufficient penetration of PDAM water in the urban area is partly attributable to the current regime of partial allocation of authority over the PDAM. In addition, due to political reasons, negotiation to solve the problem will not work. This kind of partiality problem must be solved by a higher rank in the government system, such as the provincial government or the central government. The authors conclude that the problem of partial authority allocation between regions is a matter of *externalities* arising from unique attributes of the case that we assume for depicting the institutional environment in the water supply system in Bandung region.

1.4 INSTITUTIONS OF COMMUNITY-BASED WATER MANAGEMENT

Though it is a simple voluntary association or organization, an institution of community-based water management represents a self-help group, as defined by Tao (1999), whereby through mutual understanding among members they have the ability to mobilize their local water resources to supply the needs of the members. Moreover, this institution is able to fulfil the three characteristics of the community as a group as defined by Aoki (2009), including (i) relative stability in membership, (ii) the ability to create internal rules as community norms, and (iii) the ability to utilize the norms as self-enforceable without third-party involvement. In the case of Indonesia, around 10% of citizens have access to a piped water supply system managed by the local community, generally in rural areas (BPS, 2005). In this sense, Ari et al. assume that this institution of community-based water management is a strong alternative to the public sector's existing water supply system.

Ari et al., in "A Network Analysis of Community-Based Water Management," scrutinize the social structure in the community that possesses the ability to develop a community-managed water supply system, by measuring three indices of centrality of the households through their affiliation into four types of community groups. This research found three different types of social structure that might be influenced by the strong bonding social capital formed by the community-managed water supply system, in rural Indonesia known as HIPPAM. Moreover, this approach may provide evidence of a linkage between collective action theories and the social capital approach. In addition, joint participation in a kind of community-managed water supply system such as HIPPAM might strengthen social ties among members, helping them to form and maintain trustworthiness among themselves at a steady, high level. Their ability to establish and maintain HIPPAM to pursue the common goal of having access to a piped water supply system also boosts the positiveness of social capital among villagers, enabling them to maintain trustworthiness and respect toward their living environment.

In order to investigate the mechanism of the spontaneous collaboration to access water, Ogi et al. focus on a discrete-choice model for whether a household joins HIPPAM, conditional on that household's characteristics. Economic theory suggests that the decision to join is made primarily to maximize the discounted value of future profits, so it assumes that the choice of whether to join HIPPAM or not is the result of a household's decision to maximize their utility. This research applies a spatial probit model with posterior distribution from the Bayes theorem and expresses the MCMC sampling method. The dependent variable is set to 1 for respondents who participate in CBWM (HIPPAM's members and those who get water from public hydrants), and 0 for respondents who do not participate in CBWM or for those obtaining clean water from different access (PDAM, individual wells and others). From the estimation results, it might be concluded that the length of time living in the area and the price of monthly water usage are of significance to households making decisions about whether to join a community-based water supply system.

1.5 CASE STUDIES

Over-exploitation of natural springs causes natural resources to decrease, a problem made worse by a lack of improved conservation action. Moreover, programs to conserve water springs usually rely on a biophysical approach (landscape), with no consideration of the social aspects, wherein community involvement is necessary. Yudono et al., in "A Comprehensive Social Participatory Model for Water Springs Conservation Management in Indonesia," explore the efforts of the community in Malang Regency, Indonesia. The authors recommend the legal establishment of Water Spring User Forums to cultivate the ability to manage the water springs in a certain area sustainably. The study shows that users' direct involvement in the preparation of management plans has a positive impact on water springs conservation programs. In conclusion, a sense of belonging to the conservation program of the community will strengthen community members' conservation skills, and at the same time, it will lessen their dependence on other parties.

Issues of water and social capital are associated not only with the provision of clean water but also with natural disaster mitigation, since rainfalls often create floods in most Indonesian cities. Surjono et al., in "A Cognitive Analysis of Residents and Community-Based Flood Management" investigate the social capital of three shareholders with regard to flood hazard mitigation in Indonesia: local residents, local government, and school communities. The capacity index of flood hazard preparedness was evaluated using 5 variables, based on UNESCO-ISDR and LIPI (Indonesian Science Institute) guidance: knowledge and attitude (KA), policy statement (PS), emergency planning (EP), warning system (WS) and resource mobilization capacity (RMC). As a case study, it looked at the experience of the city of Pasuruan, East Java Province. Pasuruan is representative of medium-size cities located in coastal areas and threatened by floods every year. The assessment found that Pasuruan was inadequately prepared, and that this was caused by capacity weaknesses of the local residents, school communities, and local government. To some extent, the index of preparedness in flood mitigation management represents social capital within the community because it shows the linkage of agents and norms. As argued by Murphy (2007), disaster management at the local level needs interdependency among social units although they are separate aspects of management. Conditions in Pasuruan reflect the actual situation in Indonesia, where the community is too dependent on the government. Large sections of the society, represented by the index of residents (R), have lacked understanding of the risks posed by floods and the methods of protecting against them. There were significant gaps between the residents and the government. This showed that development at the local level has not created bridging social capital within the community. The knowledge that has been gained by the government and school community has not been disseminated to local residents. Such knowledge affects how a community manages emergency planning, warning systems, and the mobilization of resources. The authors recommend that empowering and strengthening be focused on the weakest sub-variables at the local resident level, after which there should be a strengthening of capacity and connectivity among the three shareholders.

Aggraeni, in "A Case Study of Adaptive Planning and Conflict Resolution in Indonesia," attempts to formulate a strategy for resolving conflicts related to water resource utilization and adaptation to climate change. As a case study, it looked at the conflict between users and stakeholders of the Krabyakan Spring water in Sumber

Ngepoh, Lawang, Malang Regency. The evaluation found that climate change impacts on water resources availability have significant correlation. To some extent, there have been influences on water demand and food security and also on the conflict that has occurred between several beneficiaries of water resources in the study area. Finally, the author suggests six strategies: (i) improving water distribution management, (ii) strengthening the bargaining position of the beneficiaries' society, (iii) formulating climate change policy and programs that are responsive to dynamic conflict, (iv) drawing upon traditional adaptation and coping mechanisms when formulating policy, and (vi) integrating climate change in development policy and strategy.

1.6 CONCLUDING COMMENTS

This collection of papers is not meant to be comprehensive but rather to reflect diversity in the implementation (i.e., policies and strategies) of CBWM, based on some of the original ideas generated by collective action principles. We hope that the collection of papers in this volume will provide useful background as well as a stimulus to pursue future research on community based-water management and social capital.

1.7 REFERENCES

Statistic Centre Board Indonesia (BPS), 2005, Badan Pusat Statistik, Jakarta.

Devine J. (2006). Community-based organizations – new fad or old hat? *Community Development Journal*, **41**(4), 521–527.

UNICEF (2007). Global Environmental Outlook GEO 4: Environment for Development, United Nations Environrmnt Programe, Malta.

Westlund H. (2006). Social Capital in the Knowledge Economy: Theory and Empirics. Springer, Berlin, Heidelberg, New York.

2 An alternative clean water supply system for community living in coastal and flood-prone areas: lesson learned from Legon Kulon village

Ibnu Syabri

School of Architecture, Planning and Policy Development, Institute of Technology Bandung, Ganesha 10, Bandung, Indonesia 40132. E-mail: syabri@pl.itb.ac.id

Abstract

This paper presents a lesson learned from a community participation program of developing an alternative water supply system for communities in coastal and flood-prone areas. The study explores community responses to a new technology for an alternative water supply system. The factors being investigated that may influence the performance of the water supply system include community motivation, community capacity to operate and maintain a new water supply system, and ongoing support from organizations outside the community. The study provides a lesson learned of how these factors affect rural water supply system performance.

Keywords: water supply system, rural poor community, North Java.

2.1 INTRODUCTION

In this paper we describe the ongoing community services research project of the Research Center for Infrastructure and Regional Development, Institute of Technology Bandung (CIRD ITB). The project was started in 2008 and is expected to be completed in 2010. The project is a collaboration between CIRD ITB and the local government of Subang Regency and the provincial government of West Java. As part of the obligation of the center for community services, we worked with residents living in remote and poor neighborhoods. We targeted small communities on the remote coast north of Subang Regency, West Java, recognizing that they need a strategic approach to improving water services and management.

We document the project experience of creating locally-driven water management plans in Legon Kulon village, a poor and remote coastal village settlement in the northern area of Subang Regency, West Java, Indonesia. The aim of this project is to enable poor and remote neighborhoods and communities to grow well by securing reliable and well-managed water supplies. We explore the practical possibilities available, from viewing a water supply as a resource for communities pursuing their own strategic objectives, through building a small-scale water supply system, and applying the practicalities of management to site-specific issues and contexts.

Remote and poor neighborhoods, when compared with more centralized settlements, appear to have less access to basic services such as clean water and good sanitation systems, with consequent health outcomes. These remote neighborhoods generally have worse access to traditional food sources, as well as unimproved living conditions, environmentally, socially and psychologically. Overall, these communities incur high costs for water supply due to poor economies of scale, generally low levels of access to support through service networks, and limited information about different strategies (including low technology options) for securing their existing water supplies. The majority of water supplies for these communities are sourced from groundwater.

The rest of the paper is organized as follows. In the following section we briefly outline the history and current situation of water management in the Lego Kulon communities. In section three we briefly describe the principles of the frameworks used during the project. In section four we outline the process used to develop water management plans in Legon Kulon neighborhood. In section five we conclude our ongoing project by describing the lesson learned and discussing how this case study might be applied to other similar areas in Indonesia.

2.2 THE COMMUNITY AND WATER SUPPLY MANAGEMENT IN LEGON KULON VILLAGE

The focus of this study is Legon Kulon Village, Subang, West Java (Figure 2.1). The village is a small community of poverty, with 97 permanent households. The average per capita annual income in Legon Kulon Village is around 3.5 Million Rupiah (US$350) (BPS, Susenas, 2005), which is one of the lowest in West Java. The largest concentrations of the poor are in the informal settlements near coastal areas. Physically these settlements are characterized by semi-permanent structures of a single storey, a structure of normal height with walls made of bricks and the roof made of any material other than cement or concrete. Since the great tsunami in December 2005 hit Aceh and Nias, according to the local authority this area has also suffered continuous flooding due to the rise of sea level damaging road access and houses, destroying paddy plantations, and most importantly discontinuing the daily water supply delivered and carried by trucks of the local state-owned enterprise. Since that catastrophic event, the access to and supply of clean water have been limited. With regard to modes of water services for the village, no pipe networks exist; people usually use ground water, either from deep ground water or from shallow wells. Another type of distribution is by means of tank car to serve the village. Recent development indicates that there are water refill stations developed by small-scale water providers that produce drinking water to refill bottled potable water to satisfy customers' needs.

Current water supply management in this neighborhood is characterized by the lack of clearly defined responsibilities and decision-making processes among local government and support agencies. This situation leaves people in this area with little understanding of how to improve or plan water management in a strategic way, and little capacity building or agency support. However, it is important to note that most remote communities have a resource agency (i.e., local authority of sub-districts in Subang Regency) with some role in maintaining the water supply. Due to the limited capacity and skills of the local officials, roles and responsibilities vary from place to place, but may range from no record at all about the social and economic characteristics of the communities to a complete set of records that contain water quantity and quality data.

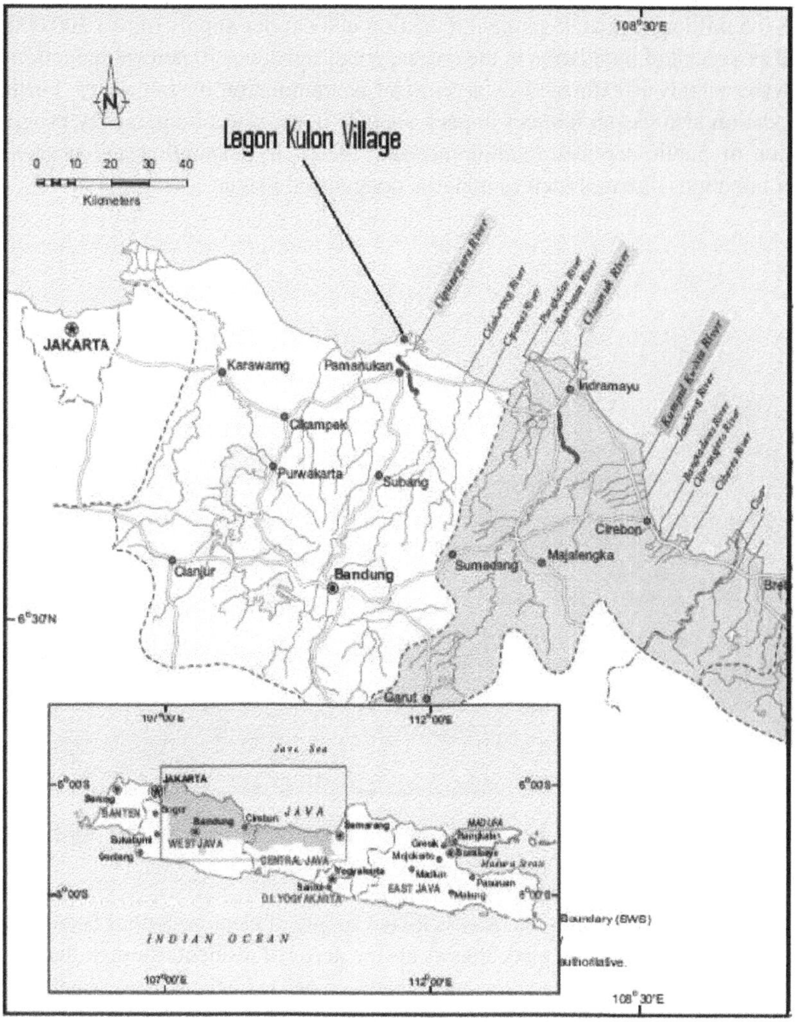

Figure 2.1 The study area, Legon Kulon village.

Several studies related to the impact of sea level rise in Indonesia (Kombaitan, 2001; Sutanta, 2002; Pinem, 2002; Kobayashi, 2003) have noted that flood inundation by sea level rise is a major threat to communities living on the north coast of Java. Flood from the sea can cause the overflow of flood-defenses like dikes, inundating thousands of hectares of paddy fields and residential houses, and more importantly disrupting water supply service. In the inundated area, several buildings have been abandoned and infrastructure has been disrupted, such as roads, local factories, and commercial establishments. Figure 2.2 shows an inundated neighborhood in Legon Kulon village.

The total inundated areas of the north coast of Subang Regency, West Java, now cover about 2000 ha extending up to 3 km from the coast. According to the local authority report (BPS-Subang, 2007), sea level rise has affected agricultural production in this area. In the Subang district alone, the loss of fish production is estimated at over 4,000 tons (valued at over US$ 0.5 million), about 95 percent reduction in local rice supply (down 300,000 tons) is estimated as a result of inundation in the coastal zone, and about 70 percent reduction in clean surface water supply is estimated as a result of contamination by salt water. Furthermore, tidal inundation also has an indirect impact, such as in the cost of emergency programs, the interruption of public services (communication, electricity, sanitation and water) and the danger of infectious diseases such as malaria, dengue and plague.

Figure 2.2 The impact of the sea level rise in Legon Kulon village, the north coast of Subang regency on houses inundated (reaching up to 50 cm).

Since that catastrophic event, the access to and supply of clean water has been limited, and a simple system has been put in place that meets the needs of residents through the use of hand pumps, for example, and harvested rainwater which is piped to kitchens. Bore water pumped (by hand) to the surface is used for all purposes other than drinking, such as for washing and bathing.

In late 2007, as shown in Figure 2.3, the local authority (i.e., the community leaders, members of Legon Kulon community, and local government officials), and the ITB project team set up a meeting to assess the problem of water supply. In the meeting, the representative of the village, local government officials and the ITB research team recognized the importance of preparing activities for future development. These included consideration of the development of water supply and management, the risk of flooding, the financial costs of floods, the risk to lives, and the physical damage to property.

The local authority and the ITB research team formed a flood appraisal group. Its aim is to promote and to enhance understanding of all issues related to flooding and water supply. The accessibility of sustainable water supplies for poor and remote neighborhoods was reviewed

Figure 2.3 A meeting in the sub-district community hall attended by community leaders, government officials, and members of the ITB research team.

by this group in 2007. In its review, the group found issues that impact the provision of reliable water supplies. These include the capacity and willingness of communities to support interventions both physically and financially in the short and long terms; and the relative lack of specialized services in the majority of poor remote communities and the difficulty in accessing specialist expertise.

Moreover, the ITB research team is responsible for developing a new system for water supply, assessing flood risk, and providing advice to the local planning authority, Subang Regency. The team can therefore advise in some cases on proposals for development of land at risk of clean water shortage and flooding. In addition to developing the new water supply system, some recommendations for the village's future development include (i) directing the development away from areas at risk of flooding, such as flood plains; (ii) identifying areas at risk wherever possible and setting appropriate criteria; and (iii) advising the development such that it shall be sited and designed to protect the natural heritage value and water quality of ponds, watercourses, aquifers and wetland within the context of water course management.

2.3 THE FRAMEWORK

There have been many frameworks for water resource and risk management and some are widely accepted as guiding principles for water supply management. For example, the World Health Organization develops water safety plans based on risk management principles for the management of all types of water supply systems, and the International Water Association has developed the Bonn Charter for Safe Drinking Water, which endorses integrated water safety plans based on risk management principles. These two frameworks outline a preventative approach to managing risks associated with drinking water. They provide direction for the design and implementation of comprehensive quality management systems from catchment to tap. This includes 'holistic guidance on a range of issues by documenting good practice for system management' (Rizak *et al.* 2003). In addition, the most important key components in these two frameworks include (i) commitment to drinking water quality management; (ii) system analysis and systems management; and (iii) supporting requirements.

Inspired by these two frameworks, the work of Grey-Gardner (2008) and Integrated Water Resources Management (IWRM) Indonesia guidelines, we developed a framework for the management of drinking water quality, primarily to address any water quality and general supply issues such as maintaining water system integrity. Using this framework, we assessed other factors, such as water needs and people's capacity to maintain a water supply; explored the combination of the water risk management approach with the livelihoods approach as a means of creating an incentive for communities to continue to manage their water supply beyond the completion of the project; and used participatory methods to build residents' understanding of water supply and management activities.

During the project, the framework process provided the foundation for engaging with communities about their water supplies and management options. The basis of this approach, as Rizak *et al.* (2003) and Cunliffe (2005) suggested, is to promote an understanding of the entire water supply system, identify the potential barriers, apply multiple barriers and establish critical control points to reduce exposure to hazards.

We employed throughout this project a fundamental principle suggested by Grey-Gardner & Walker (2002), which is the embedding of capacity-building initiatives. Based on this principle, the process to engage participating communities is necessarily designed to be flexible and adaptive to suit the purposes and circumstances of the local people. We spent the first three months establishing the case study settlement sites and doing background research. During this phase, we also identified and negotiated stakeholder roles and project objectives.

As shown in Figure 2.4, our research team worked together with the community in constructing the groundwater well. The concept of working with communities that have a purpose or a livelihood was central to our selection of the case study community. The premise was that a livelihood would provide the incentive to manage a settlement water supply, since

Figure 2.4 Constructing the groundwater well.

the water supply would be required for the continued functioning of the community's purpose, but the responsibility rests mainly with the residents. Some criteria employed for selecting the neighborhood are a permanent population of less than 100 households; primary responsibility for the water supply falling on residents; a strong livelihood or purpose; and a willingness to be a part of the project and a commitment to making decisions about how to better manage the water supply.

In the participatory process, the Sustainable Livelihoods assets pentagon was used as a tool to engage and enable residents' aspirations to be part of the planning process. The assets pentagon was completed with participants twice during the project – during the early identification of residents and of a settlement's assets, and as an evaluation tool to identify any changes during the project. When using the Sustainable Livelihoods worksheets, participants were asked to identify, rank and analyze problems and opportunities as they perceived them.

2.4 WATER MANAGEMENT PLAN

During the meetings about future water management, participants were keen to participate in the training program, which was seen as an opportunity to work together and make some significant improvements to the property's water supply. It was also considered an important opportunity to clarify management roles and responsibilities for tools and equipment. Rectifications to the neighborhood water supply were identified during the development of the management plan. The project participants tried to make the recommended improvements themselves, including constructing the groundwater well, before the training program started.

The residents wanted a water supply manual that they could keep. The manual would include information about tasks to maintain the water supply system, and would thus help with maintenance and financial planning.

The training program in the neighborhood had five participants. It was seen as an opportunity to share knowledge with the residents at Legon Kulon and work together to make immediate rectifications to secure the water supply. The training activities were based on the water manuals and included the following:

- building a groundwater well
- building a fence around the well
- building pipes, including installing a new tank for tapping water
- Understanding the technical design of the windmill pump
- Understanding the technical operation, troubleshooting and maintenance of the windmill pump

The training program in the neighborhood focused on increasing the skills of residents in identifying and understanding risks to the water supply. Skills were built to make sure residents could solve problems and maintain the water supply themselves. To supplement the learning, residents also learned how to secure the water supply.

In this project we built a windmill (see Figure 2.5), used to pump water from the well as an alternative system of water sources. Although this technology is not new to the community, the building and operation of this technology may require more highly skilled manpower. One main question that we wanted to test was how sensitively modern technology affected the level of participation. In this case we realized that technological choices must take account of specific conditions prevailing at the location of use. This means that the most advanced and

Figure 2.5 The windmill as an alternative for water supply.

modern technology is not necessarily the optimal choice in all cases. If the system cannot be sustained because of a lack of spare parts, skilled manpower or economic resources for operation, it is not the most appropriate solution. Moreover, high-cost technologies may prevent community and household involvement in water management.

2.5 CONCLUSION

The project was evaluated on 26 January 2008. We used the Sustainable Livelihoods assets pentagon to assess changes in the residents' capacity for managing their own water supply after the project was completed. The residents showed a solid understanding of problems and issues related to the water supply and a realistic understanding of its limitations. Although the participants have not had the skills to make basic repairs and maintenance yet, more people in the neighborhood are now interested in participating in the project. The most significant change in this neighborhood is a new understanding of the responsibilities associated with water management and the willingness to take responsibility for regular maintenance activities.

2.6 REFERENCES

BPS-Biro Pusat Statistik (2008). *Data Potensi Penduduk Indonesia*. Biro Pusat Statistik, Jakarta, Indonesia.

BPS-Subang (2007). Subang dalam Angka. Biro Pusat Statistik Subang, Subang, Jawa Barat.

Cunliffe D. (2005). Management of water quality in rural and remote communities, NHMRC initiatives in rural health workshop, *Central to Health: sustaining well-being in remote and rural Australia*, paper presented at 8th National Rural Health Conference, 10–13 March.

Grey-Gardner R. and Walker B. (2002). What Lies Beneath: Sustainable Groundwater Management for Communities of Indigenous People, *Proceedings of the International Groundwater Conference, 'Balancing the Groundwater Budget'*, 2002, Darwin, International Association of Hydrogeologists.

Grey-Gardner R. (2008). *Remote Community Water Management*, DKCRC Research Report 27, Desert Knowledge Cooperative Research Centre, Alice Springs.

Kobayashi H. (2003). *Vulnerability assessment and adaptation strategy to sea-level rise in Indonesian coastal urban areas,* National Institute for Land and Infrastructure Management, Ministry of Land, Infrastructure and Transport, Asahi-1, Tsukuba-city, Japan.

Kombaitan (2001). Sustainable urban coastal planning: searching of frame analysis for the mitigation of the impact of sea level rise (In Indonesia). In: *Seminar Proceeding of the impact of sea level rise in Indonesian urban environment*, Public Works Department - Indonesia and Building Research Institute, Bandung.

Pinem F. (2002). *The use of aerial photographs for susceptibility mapping of flood hazard in Kali Garang Semarang Central Java (In Indonesian).* Undergraduate thesis, Geography Faculty, Gadjah Mada University, Yogyakarta, Indonesia.

Rizak S., Cunliffe D., Sinclair M., Vulcano R., Howard J., Hrudy S. and Callan P. (2003). Drinking water quality management: a holistic approach. *Water Science and Technology,* **47**(9), 31–6.

Sutanta H. (2002). *Spatial modeling of the impact of land subsidence and sea level rise in a coastal urban setting, case study: Semarang, Central Java*, Indonesia. M.Sc. thesis, International Institute for Geo-Information and Earth Observation, ITC, Enschede, The Netherlands.

3 Collective action in water resource management: theoretical perspectives and propositions

Gautam Ray

Graduate School of Management, Kyoto University, Japan. E-mail: gautamray79@gmail.com

Abstract

This paper reviews the theoretical perspectives in Elinor Ostrom and Robert Wade's frameworks of collective action in the management of CPR including water resources and brings out the inadequacies in the extant literature for explaining collective action in diverse settings. It then introduces a general theory of collective action and discusses how the general theory can be applied in the context of collective action in the management of water resources. It posits structuration and legitimation dynamics that can sustain collective action over a period of time. A set of propositions is deduced from the general theory for their empirical validation.

Keywords: collective action problem, free rider problem, Prisoner-dilemma, focal actor, potential actor, shared mental model, endogenous preference, structuration and legitimation dynamics.

3.1 INTRODUCTION

Managing a constant if not depleting level of available water resources has been a challenging problem worldwide for a number of reasons. First, competing demands on common water resources from households, agricultural firms, industries, fisheries, and the producers of energy, ecological and transportation services have been generating conflicts of interest among these different users with different natures and levels of demand for water. The problem is compounded by the fact that water is a non-substitutable essential good that is absolutely necessary not only for maintaining day to day life but also for producing life sustaining basic goods and utility services.

Conflicts over water resources have been especially pervasive between the large landowning class versus small and marginal farmers, industry versus households, and upstream users versus tail-end users of river basins, canals and watersheds. For example, Coca-Cola plant's large-scale use of groundwater has been sparking public protests from time to time in different parts of the world. In fact, the Coca-Cola plant in the state of Kerala in India had to be shut down in 2006 as residents of Plachimada village rose in protest over the scarcity of water for their household and agricultural uses, which they believed to be caused by the plant's excessive pumping of groundwater resource.

The second major reason is asymmetry in the access to and power over common water resources. Due to this asymmetry, users having higher access tend to over-consume and waste water to the detriment of the interests of other users. Very often downstream users of watersheds, canals and river basins find themselves victims of over-consumption and wastages by upstream users, particularly in lean supply seasons. Ironically, the same downstream users also find their homes and agricultural land flooded in high rainfall seasons.

The problem of inequity in access is aggravated by the rationality paradigm, which gives upstream users a sense of legitimacy to consume as much as they can in utter disregard of the severe distress caused to tail-end users by their unrestrained consumption. As shown by many researchers, including Wade (1982) and Bardhan (1993), rich landlords and other major appropriators collude with state agencies to systematize their overexploitation of the commons and obstruct changes in existing institutional arrangement that could remedy the prevailing structure of exploitation. In the case of trans-boundary river basins, the asymmetry in access to water resources in international river systems, aggravated by changes in the hydrological cycle, leads to the threat of violent conflicts among nations.[1]

A third important reason is that water resources tend to be grossly undervalued by the society because of the age-old social norm and belief that the state should supply water to households and the agricultural sector at a low price, if not entirely free, since it is a naturally regenerating resource and is essential for sustaining life. The economic pricing of water, crucially important for managing the demand on the overstressed water resource, is generally perceived to be inconsistent with the prevailing social norm, value, belief and expectation about the state's duty to supply water at a low price. Consequently, municipalities and irrigation departments are constrained to supply water at highly subsidized prices even to households and firms who can afford to pay the marginal cost of its production. Already faced with severe budget constraints, most municipalities then end up undersupplying water in slums and poor neighborhoods, further impoverishing the poor. These socio-political dynamics of water-market failure, quite rampant in developing economies, illustrate why most households do not get a 24×7 supply of water and why most people do not have access to safe drinking water.

Even though rapid industrialization and high economic and population growth have increased demands on available water resources and such high demands have aggravated the scarcity, most developing economies find it difficult to increase the price of water to the level of its incremental marginal cost. As a result of this social undervaluation, it is difficult to manage the level of water's demand and supply effectively or efficiently through market pricing mechanism. Even if property rights could be assigned to this overstressed resource, it is difficult to price water at a progressively higher price when it is consumed beyond the level (say, 20 liters per day per person) that is considered to be necessary for day-to-day living by international standards.

Managing common property water resources, namely groundwater aquifers, watersheds, lakes, ponds, and river basins, poses special challenges arising from their quasi-public good property, that is, their attributes of non-excludability and some sort of rivalry in consumption. Non-excludability of such resources makes it difficult for a regulatory authority to restrict

[1]Before the Six-Day War between Israel and its Arab neighbours in 1967, Prime Minister Levi Eshkol had declared that "water is a question of life for Israel" and therefore "Israel would act to ensure that waters continue to flow" (Gleick 1993:85). After signing the peace treaty with Israel in 1979, Egypt's President Anwar Sadat observed, "The only matter that could take Egypt to war again is water" (Starr 1990:19).

their aggregate consumption to a level equal to or below the physical or biological rate of their regeneration, a problem that is already severe in populous developing countries such as India.

This race for the over-use of common water resources is generally perceived as 'rational' by consumers who have the access and capacity to over-consume. In their calculus, the net benefit of their choice of 'more consumption now' exceeds their share of the eventual cost of a depleted natural resource that will be divided among multitudes of present and future consumers. In such a game, 'more consumption now' is the dominant strategy giving rise to the Prisoner Dilemma (PD) game-type suboptimal outcome, which leads rational appropriators to the race for destructive overconsumption of commons, a phenomenon Garett Hardin referred to as 'the tragedy of the commons' in his classic 1968 work published in the Science journal.

In principle, the 'Hardinian tragedy' can be mitigated through state regulation restricting over-consumption of common resources, provided the regulators are capable of enforcing the restriction and have the incentives to do so. But the empirical reality, as captured in Wade (1982) in the context of the irrigation system in India, is that state agencies are often the perpetrators of systematized violation of statutory rules and restrictions on water use, and this system is greased by institutionalized corruption flowing from the top political masters down to the lowest functionaries in the irrigation department of state governments.

In the four decades since Robert Wade's study of canal irrigation management in south India, all that has changed is the size of corruption, which is now magnified in proportion to the multiple-fold increase in the stakes involved in over-appropriating common water resources. In the management of common water resources in India and pretty much across the world, government failure still remains as pervasive as market failure is. This leaves collective action by the community of users as the only possible mechanism for bringing about a superior outcome in the provision, maintenance and appropriation of common water resources.

Collective action literature, however, discusses several problems in organizing, operating and sustaining collective action in the management of common water resources. At the theoretical level, three models capture collective action problems (CAPs): Merill Flood and Melvin Drescher's Prisoner Dilemma (PD), originally developed as a two-person one-shot game in 1950; Mancur Olson's *The Logic of Collective Action* (1965); and Garett Hardin's "The Tragedy of Commons" (1968). The empirical reality is that although collective action problems have been resolved successfully in many cases as shown by game theorists and empirical investigators, cases of failures of collective action are more abundant.[2]

Summarizing his research findings on the management of common water resources in South India, Wade (1987) observed: "Within one small area, I found some villages with a more sophisticated set of arrangements for canal irrigation and grazing (and for supplying other public good as well) than had been previously reported for Indian villages. Other villages nearby, however, showed no sign of these arrangements. The dismal frequency of degraded grazing commons, despoiled forests, over-exploited groundwater and depleted fisheries shows only too clearly that collective action cannot be presumed to be always a viable route. On the other hand, there are no good analytical reasons for presuming that it will generally fail" (pp. 95–96).

[2]In fact, persistent problems of systematic destruction of environmental/natural resources, hunger, malnutrition, grossly inequitable access to opportunities for human development, and extremely perverse income distribution are stark examples of not only market failure and government failure but also collective action failure at all levels of society.

While appropriation of common pool resources (CPRs) is usually beset with problems such as those captured by the prisoner dilemma type suboptimal outcome in the short-medium run and the 'Tragedy' in the long run, provision and maintenance of CPRs typically face 'Olsonial Logic' of free riding behavior by potential beneficiaries, each one of whom wants to minimize his or her effort, hoping that others will do all that is necessary to provide for the common resource. If all potential beneficiaries are rational, then nobody will invest enough effort to maintain or provide the resource. Thus, "unless the number of individuals in a group is quite small, or unless there is coercion or some other special device to make individuals act in their common interest, rational self-interested individuals will not act to achieve their common or group interest" (Olson, 1965).

Elinor Ostrom, a political scientist who won the Nobel Prize in economics in 2009 for her research on governance of CPRs, dedicated her lifetime research to resolving the collective action paradox: individually rational strategies lead to collectively irrational outcomes. She believed almost as a matter of faith that rational human beings can achieve rational results. In pursuit of her belief she examined successful as well as failed cases of participative management of CPRs. She also studied the experimental games in public good provision conducted in laboratories in order to identify a set of contextual variables and institutional rules that explain the resolution of collective action problems in the provision of public goods.

Ostrom (1990, 2000) and Wade (1982, 1987) opened up a line of research inquiries that sought to apply theoretical insights captured from the critiques of the three CAP models to field studies in order to identify the conditions and institutional regimes under which communities can successfully organize participative management of common resources. These research results showed how in some settings the potential beneficiaries could organize collective action in water resource management and sustain it by coping with free riding, solving commitment problems, arranging for the supply of new institutions, and effectively monitoring individual compliance with the help of a set of rules.

The fundamental precept in this body of literature is that the real challenge in managing the commons lies in the structure of prevailing situations, which is responsible for the mal-governance of the common resources. This structure includes: the taken-for-granted legitimacy in maximizing the appropriation of common resources which belong to everybody and therefore nobody in particular, inadequate regulatory oversight, nexus between appropriators and regulatory agencies and the prevailing social norm of tolerating such misappropriation. The common refrain is that appropriate institutional arrangements legitimized by users' participation in the design and implementation of 'rules in use' can change the structure of situations.

A big question, however, remains largely unanswered: how societies could bring about legitimate institutional arrangements that could sustain collective actions of all sorts in all settings, preserve humanity's stock of precious common resources such as water and other ecological resources, and effectively manage the flow of utilities from such resources for the present as well as future generations? So far the dominant approach to answering this question has been to extract inductive propositions based on a small sample of success and failure stories. But this body of literature has yet to throw adequate light on a number of fundamental questions on collective action

To quote Elinor Ostrom: "We need to understand how institutional, cultural and biophysical contexts affect the type of individuals who are recruited into and leave particular types of collective action situations, the kind of information that is made available about past actions, and how individuals can themselves change structural variables so as to enhance the probabilities of norm-using types being involved and growing in strength over time. Further developments along these lines are essential for the development of public policies that enhance socially beneficial, cooperative behavior based in part on social norms" (Ostrom, 2000; p. 154).

A first step in this direction would be to examine analytically the theoretical foundations of a generalized collective action. The second step is to apply the general theory of collective action in the particular case of a collective action, say, in the management of common water resources, and then extract deductive propositions from the general theory as applied in a particular case for the purpose of their empirical validation. The third step is to refine the propositions and/or the analytical frameworks in the light of empirical findings. The final step is to take a relook at the general theoretical framework and refine it, if necessary, with the help of empirical results and bring up new theoretical propositions for empirical validation.

This paper takes the first two steps: it develops the general theory of collective action introduced in Ray (2013) and applies the theory in the case of management of common water resources. Before proceeding to take these two steps, it is necessary to define collective action. Broadly defined, collective action (CA) refers to any jointly performed act, including the provision and maintenance of public and quasi-public goods for the common benefit of the actors.[3] For the purposes of this paper, however, CA refers to a range of actions that potential beneficiaries (say, water users) voluntarily take with the help of supporting organizations (say, NPOs and the state) in order to remedy the prevailing unsustainable, unjust and inequitable appropriation of the commons (say, common water resources).

The rest of the paper is organized as follows. In the next section, Elinor Ostrom's framework of collective action is discussed, which is followed by a brief explanatory discussion of Robert Wade's narrative of the conditions for successful collective action. In Section 4, a general theory of collective action introduced in Ray (2013) is further developed. Section 5 discusses the role of 'structuration' and legitimation dynamics in sustaining organized collective action. In Section 6, the general theory is applied in the case of water resource management and a set of theoretical propositions is deduced for empirical validation. Section 7 concludes the discussions.

3.2 COLLECTIVE ACTION FOR MANAGING COMMON POOL RESOURCE (CPR): ELINOR OSTROM FRAMEWORK

In her work *Governing the Commons: The Evolution of Institutions for Collective Action*, Nobel Prize winner Elinor Ostrom identified similarities in the institutional arrangements of long-enduring common pool resources (CPRs) governed by beneficiaries in Swiss alpine

[3]To be sure, all collective actions are not necessarily beneficial to society. Collective exploitation of landless laborers (serfs) by the rich and powerful (through the institution of bonded labor) is one example of structural violence that is still manifest in different forms. Sometimes collective actions can actually harm the group of actors or their community. Violent racial and religious conflicts between Serbs and Croats, Serbs and Muslims, Croats and Muslims, Hutus and Tutsis, Sunnis and Shias, and Northern Irish Catholics and the British-Northern Irish Protestants have been raising serious doubts about humanity's ability to co-exist peacefully and work towards common prosperity.

meadows in Torbel, Japanese mountain commons in Hirano, Nagaike and Yamanoka villages, local irrigation systems such as Spanish huertas in Valencia, Murcia, Orihuela, and Alicante, and Philippine zanjeras. She also analyzed institutional changes in a series of California groundwater basins and identified a set of 'internal characteristics' of community of CPR appropriators who would be more likely to supply a 'series of incremental changes in operational rules to improve joint welfare'.

The following are the internal characteristics: 1) most appropriators share the common view that they will be harmed if they do not adopt an alternative rule; 2) most appropriators will be affected in similar ways by proposed rule changes; 3) they highly value the continuation of activities from the CPR; in other words, they have low discount rates for future consumption; 4) they face relatively low information, transformation and enforcement costs; 5) they share generalized norms of reciprocity and trust that can be used as initial social capital; and 6) the group appropriating from CPR is relatively small and stable (ibid. p. 211).

Ostrom (2000) enumerates a number of contextual variables that have been identified by researchers as conducive or detrimental to collective action in providing and managing common/public goods. These variables are the following: i) the type of production and allocation functions; ii) the predictability of resource flows; iii) the relative scarcity of the good; iv) the size of the group involved and the heterogeneity of the group; v) the size of the total collective benefit; vi) the marginal contribution by one person to the collective good; vii) the size of the temptation to free ride; viii) the loss to cooperators when others do not cooperate; ix) a choice of whether or not to participate; x) the presence of leadership; xi) past experience and level of social capital; xii) the autonomy to make binding rules; and xiii) a wide diversity of rules that are used to change the structure of the situation.

Ostrom (2000) suggests a possible mechanism of organizing collective actions for effective management of CPRs. According to this mechanism, the first step is the formation of the nucleus of trust and reciprocity for cooperative behavior when a small number of users who are 'conditional cooperators' constitute a cohesive group with agreed-upon boundary rules for the entry of new members. The presence of a leader or entrepreneur is frequently an important stimulus (Frohlich *et al.* 1971; Varughese, 1999). Once this nucleus is formed, a set of eight design principles of CPR institutions helps in the evolutionary growth of this trust and reciprocity, which sustains collective action over generations.

These design principles are the following: i) clear boundary rules; ii) restriction of benefits to individual users by locally devised rules which are fair, just and cost effective in their enforcement; iii) inbuilt flexibility in the institutional arrangement to allow users to make new rules and modify existing ones to remove any anomalies, inequities, and hardships; iv) selection of monitors who are committed, honest and effective in overseeing the condition of the CPR and keeping an eye on its use or misuse; v) graduated sanctions for violation of rules; vi) a low-cost conflict resolution mechanism; vii) minimal recognition by the government authority of the community's right to organize and manage resources; and viii) organization of government activities in multiple layers of nested enterprise for the management of more complex CPRs that cannot be effectively or efficiently managed by communities.

Ostrom (2000) made a first pass towards developing a general theory of collective action. Her approach was to develop a theory which explains certain 'core facts' that have been replicated frequently in a number of experimental studies.[4] These facts are the following: i) in

[4]Ostrom (2000) refers to Davis and Holt, 1993; Ledyard, 1995; and Offerman, 1997, for an overview.

a one-shot public good game, as well as in the first round of finitely repeated games, subjects contribute 40% to 60% of their endowments, but the contribution level diminishes after the first round—over 70% of subjects contribute nothing in the announced last round of a finitely repeated sequence; ii) those who believe that others will cooperate are more likely to cooperate; iii) learning games better leads to more cooperation; iv) face-to-face communication produces substantial increases in cooperation that are sustained even in the last period; v) subjects are willing to spend their personal resources to punish defaulters; and vi) the rate of contribution to a public good is affected by "various contextual factors including the framing of the situation and the rules used for assigning participants, increasing competition among them, allowing communication, authorizing sanctioning mechanisms, or allocating benefits".

To explain these 'core' facts, Ostrom (2000) assumed that there are three types of individuals: conditional cooperators; willing punishers; and 'rational egoists' who are the neoclassical *homo economicus*. Conditional cooperators follow social norms and are the trustworthy type. Rational egoists do not follow social norms if those do not enhance their personal benefits. They cannot be trusted as reciprocators or cooperators. In the absence of complete information about types of individual players, Guth and Kliempt (1998) show that trustworthy types will survive as a substantial proportion of the population if there is a noisy signal about the distribution of players' types in the population. A noisy signal may result from mutual face-to-face communication, and from various mechanisms designed to monitor each other's behavior. Ostrom (2000) relies on this result to infer that face-to-face communication in the community and effective rules-in-use to discourage deviant behavior promote a cooperative outcome that cannot be explained by the conventional rational-egoist model of human behavior.

Ostrom and Ahn (2003) formulated a theory of collective action that is founded upon social capital. They posit that a set of contextual variables and trust are the precursors of collective action and that trust is generated from social capital variables, namely, norms, networks and institutions.

3.3 CONDITIONS OF COLLECTIVE ACTION: ROBERT WADE'S PERSPECTIVE

What does it take for communities of potential beneficiaries to solve a common set of problems and govern their behavior by following appropriate sets of design principles? Why can one group of communities voluntarily organize and design rule-based management of their commons while other groups of communities in the neighborhood fail to do so?

In his paper published in the *Cambridge Journal of Economics* in 1987, Wade observed: "My study examined the institutions of collective action for resource management in 31 irrigated villages in a single district of South India. I was particularly interested in why some villages have an elaborate form of organization for village-based common-poll resources management, while other villages in the sample have none; only a few miles might separate villages of each type. Such organizations are entirely independent of state. I found that the main factor explaining the presence or absence of collective organization is the net collective benefit of that action" (p. 102).

"The issue of the voluntariness of collective action therefore has to be considered at two levels. At the constitutional levels, people can negotiate voluntarily a set of rules of restrained access or financial contributions, their incentive to do so being the prospective collective benefit. At the level of action, most of the compliance with the rules must also be voluntary, not the result of a calculus of evasion and punishment. But the rules must be backed by a system of punishment, the existence of which helps to reassure any one person that if he follows the rules he will not be suckered, and which aims at times of crises to directly deter. This argument makes the size of the net collective benefit the major factor in explaining the presence or absence of corporate organization in groups like Indian villages" (p. 103).

At the end of this paper, after critiquing the three models of collective action, namely, 'Prisoner Dilemma', 'Tragedy of Commons' and 'The Logic of Collective Action,' Wade (1987) spelled out a set of conditions that increase the likelihood of successful organization of collective action by users. Although he did not discuss how these conditions can explain successful collective actions, the explanations can be surmised from his critique of the three models of collective actions and the extant literature on collective action theory.

Condition 1: The Resources

The smaller and more clearly defined the boundaries of the common pool resources, the greater the chances of success

Explanation: The theoretical explanation of this rule comes from the fact that smaller and more clearly defined boundaries assign club good attributes to a common pool resource since such boundaries make it easier for the community to exclude outsiders from its appropriation. A small common pool resource such as a tank or a pond is accessible to a small group of users who can organize collective action more easily than a large group because members of a small group may find it difficult to free ride on the effort of others. The Williamsonian notion of lower transaction costs for a small-sized common resource with clearly defined boundaries also explains why the cost of organization of such a collective action would be less than that of a bigger-sized resource.

Condition 2: The Technology

The higher the costs of exclusion technology (such as fencing), the better the chances of success

Explanation: The high cost incurred by users for technology such as fencing makes a common property resource more like a valuable club good with high membership fees. Because of their knowledge of the high potential benefit of the common resource, users incur the high cost of exclusion technology and organize collective action to eliminate free riding through such technology.

Condition 3: Relationship between resources and user group

- i) *Location: The greater the overlap between the location of the common pool resources and the residence of users, the greater the chances of success*
- ii) *Users' demands: The greater the demands (up to a limit) and the more vital the resources for survival, the greater the chances of success*
- iii) *Users' knowledge: The better their knowledge of sustainable yields, the greater the chances of success*

Explanation: i) Location condition makes the common resource a joint property of the local community rather than an open-to-all common. Free riding without getting noticed is difficult as users within a local community are well informed of one another's behavior. A PD outcome is also not possible because of the repeated use of such resource and the possibility of gathering information at low cost about others' use of the resource through face-to-face communication. Evolutionary game theoretic models such as Guth and Kliemt (1998) suggest that there will be a cooperative equilibrium outcome in such a community setting where there is likely to be a large number of' 'conditional cooperators' or trustworthy actors who will have full access to the common property resource over a long period of time.

ii) Higher demand and higher necessity for the good means a higher stake in the success of collective action arising from higher potential benefit and perceived need for the resource, which puts a correspondingly higher cost on depletion of the resource in the case of its over-appropriation.

iii) Better knowledge of sustainable yield from the resource increases the value of the resource to the users, which increases the potential benefit of restrained use of the resource. Wade's field studies showed that the success or failure of collective action is directly dependent on the net potential benefit of restrained use of the common resource.

Condition 4: User Group

i) Size: The smaller the number of users, the better the chances of success

ii) Relative power of sub-groups: The more powerful those who benefit from retaining the commons are, the better the chances of success

iii) The better developed such arrangements are among the users, the greater the chances of success

iv) Extent to which users are bound by mutual obligations: The more likely that promises will be kept, the better the chances of success

v) Punishment against rule breaking: The more the users have joint rules for purposes other than common pool resources, and the more bite behind those rules, the better the chances of success

Explanation: i) According to Olson, the collective action problem is least for small groups because of the higher cost of free riding and the lower transaction cost. This point has already been explained in Condition 1

ii) This condition for success can be explained by the Oliver-Marwell theory of provision of public goods that assumes an s-shaped production technology for public goods. As the more powerful beneficiaries (activists) having higher interest in public good invest upfront sufficiently to take the investments of non-activists to the steeper part of the production function, nobody free rides and the public good is provided even though the group size may be large.

iii) This condition for success can be explained by the fact that a higher level of social capital in the community, such as clubs and associations for social exchanges, provide a platform for communication among users for discussion of the rules for management of common pool resources.

iv) This condition can also be explained by the fact that a higher level of bonding social capital among the users makes them conditional cooperators facilitating collective action, as explained in condition 3 (i).

v) The pre-existing institutions that provide for penalty for violation of rules also create social capital of trust and reciprocity within the community, and such social capital promotes the cooperative behavior necessary for organizing collective action.

3.4 A GENERAL THEORY OF COLLECTIVE ACTION

Ray (2013) introduces a general theory of collective action which posits that all collective actions are founded upon four structural factors: i) endogenous preference and behavioral choice of individual actors; ii) 'shared mental model' or shared knowledge and its accessibility among actors; iii) 'focal actor(s)' and 'potential actors'[5] of collective action; and iv) prevailing objective conditions in the society that facilitate or constrain collective action. It is argued that all collective actions can be explained in terms of focal actors' human and shared knowledge network, its accessibility among potential actors and the prevailing objective conditions in the society, including social capital resources accessible to potential actors.

3.4.1 Individual Preference and Choice

The general theory is based on the assumption that individuals have a two-dimensional preference structure which evolves endogenously through their learning experiences. On one dimension, individuals are concerned with their material well-being in which their self-interests[6] dominate, while on the other dimension they are concerned with their perceived or assumed well-being based on non-material influences unrelated to narrow self-interests. Such perceived or assumed well-being is rooted in individuals' values and a normative sense of duty, commitment, justice, fairness and equity that are learned from their social milieus and cultural networks.

While the self-interest dimension of human preference comes from the 'animal spirit' of *homo economicus*, preference for non-material well-being, although originally sourced in individuals' intrinsic or core values, is also socially and culturally constructed from the prevailing influential ideas, norms, symbols and ideologies in their social and cultural milieus. Behavioral choices of the same individual can be different depending on which dimension of his preference structure dominates in a given context.

In the presence of uncertainty associated with complex problems arising from changing socio-economic space, an individual may not know *ex-ante* whether his self-centered behavioral choice may be actually beneficial to him *ex-post*. In such a setting even a predominantly selfish individual may draw upon his otherwise weaker preference for non-material well-being and choose to cooperate. In the course of this newly developed proclivity

---

[5]The adjective 'focal' is used advisedly to qualify those actors of a collective action who are focused on the social need for such an action and not on their personal benefits. The notion of focal actors is similar to that of activists in Oliver and Marwell (1988 and 1992). Both have high levels of motivation, interest and resources to mobilize collective action. The subtle difference between the two notions is that focal actors are driven purely by their inner call or 'shared mental model' (Denzau and North, 1994), while activists are more like entrepreneurs who 'have to worry about costs and benefits' of collective action (Oliver and Marwell, 1992).

Potential actors, on the other hand, are more like Oliver-Marwell's notion of 'non-activists', who have less resource endowment, interest and knowledge about collective action but are willing to throw in their lot in collective action only if they access and internalize focal actors' shared knowledge or mental model and start to believe that the collective action is consistent with their 'enlightened self interest'.

[6]Self-interest is treated as exogenously given in most collective action models, as if it is an innate preference which is so powerful that it always overrides other commonly observed human traits and tendencies such as compassion, altruism and sense of justice and fair play.

towards cooperative behavior he accesses shared knowledge or ideas about the potential collective benefits of cooperative behavior. This knowledge strengthens his preference for non-material well-being as he begins to identify himself with others in the community.

In this process the individual starts to incorporate 'enlightened self-interest' in his preference structure, a process that may be referred to as 'structuration'. Motivated by his endogenously formed preference for enlightened self-interest, he starts to map into his own utility function the effects of his action on the preference structure and utility of other members of the community. Through this structuration process his preference structure undergoes change from narrow self-interest seeking free riding behavior to cooperative behavior through the social learning process. Thus cooperative behavior for collective action emerges as the collective choice of individuals who can access and internalize a shared mental model or shared knowledge of the net benefit of collective action, a phenomenon that can be referred to as social learning.

As shared knowledge of the benefit of collective action diffuses in the community, more people cooperate, creating a new value system or social norm for cooperative behavior in the society. These learned behavioral changes, induced by accessing and internalizing some ideology or shared mental models and shared knowledge, are not rare at all. Individuals have been sacrificing their material well-being, even their lives, driven by a sense of justice, duty or commitment rooted in ideologies (communism, for example) or beliefs or myths or ideas (nationalism, for example), or by the desire to promote group interest or solidarity (trade unionism, for example).

Such endogenous development of preference for non-material well-being takes place through formal education and/or informal social learning in the wider socio-cultural environment (cf. Mantzavinos *et al.* 2004). Education/collective learning systems in parental/familial, institutional and wider socio-cultural environments can attach powerful meanings, symbols and influential ideas to behavioral choices driven by notions of 'duty', 'commitment' and 'equity'. They are a powerful vehicle for developing myths, dogmas, belief systems and ideologies, or what is referred to as 'shared mental models' (Denzau and North, 1994).[7] These shared mental models evolve over time as they are transmitted across generations (cf. Mantzavinos *et al.* 2004) and constitute what we call 'shared knowledge'.

By accessing the stock of shared knowledge or the underlying shared mental models, some highly perceptive individuals acquire personal values, integrity of character and the cognitive capability to further develop or refine the stock of shared knowledge, a non-rival knowledge capital, out of which their non-material preferences and choices are endogenously formed.

At what level of material well-being does an individual start to develop a stronger preference for non-material well-being? The answer depends on his intrinsic values, which in turn depend on his knowledge endowment. Men of the type of Socrates with rich knowledge endowment are likely to have a lower threshold level of material well-being, while men of the type of Genghis Khan[8] may remain insatiate despite amassing fabulous wealth and power. Of course, these are extreme examples; the point is that some men get hooked on power and material wealth while others become content after reaching a reasonable level of material

---

[7]While mental models are internal representations that individual cognitive systems create to interpret the environment, institutions are external (to the mind) mechanisms individuals create to structure and order the environment (ibid., p.4).

[8]The example of Genghis Khan is taken from Basu (2011).

wealth because of they know that they would be happier pursuing their non-material well-being rather than consuming more goods and services or accumulating more material wealth.

Such individuals who become content and satiated with a lower threshold level of material well-being pursue acquiring, developing and refining the existing stock of shared knowledge in order to promote their non-material well-being. We call such individuals 'focal actors', a notion developed in the next sub-section. These focal actors develop knowledge, or 'mental models, ideologies and institutions' (Denzau and North, 1994) and share them with common people and mobilize the 'critical mass' (Oliver and Marwell, 1988) of human and knowledge network. The actors within this network have mutual trust and reciprocity as well as just and altruistic motivations that are rooted in their common preference for non-material well-being and enlightened self-interest.

It is possible to see some similarity between this notion of a 'critical mass' of human and knowledge network and Ostrom's notion of a 'nucleus of trust', the formation of which is the first step in the organization of collective action as posited by her in Ostrom (2000). But there is a difference. Unlike Ostrom's framework, this general theory gives a micro-foundation to the formation of the critical mass of human and knowledge network—the micro-foundation being the structuration process of change in the preference structure of individuals faced with complex and uncertain problem space. Such a change occurs endogenously through the process of accessing and internalizing shared knowledge or mental models of the expected benefits of cooperative behavior for collective action.

An individual has multiple identities (Sen, 2006), which includes her group or class or social identity alongside her self-centered egotistic identity. Social, cultural and economic conditions may determine which identity becomes the dominant one in her behavioral choice. It is perfectly conceivable that the same individual behaves differently in two different milieu. Such behavioral contrasts may result from strategic interactions, as shown in many game theoretic frameworks; they may also result from endogenous change in preference from narrow self-interest to enlightened self-interest.

Assuming that production of collective action is governed by an s-shaped production function, the same individual who had earlier perceived free riding as the rational strategy would have chosen to cooperate and participate in a collective action if a 'critical mass' of focal and potential actors had invested upfront large enough resources to take the production function to its steeply increasing range (Oliver *et al.* 1988). The consequential high return from the complementary investments made by potential actors following the upfront investments of focal actors reaffirms the validity of the shared knowledge about the viability of collective action. Such reaffirmation attracts more participants in the shared knowledge network for collective action, increasing the power and productivity of the human and knowledge networks of focal and potential actors.

All collective action problems may thus be resolved if a group of actors endowed with shared knowledge or mental models act upon their preference for non-material well-being and build a network of people having similar preference structure and shared mental models. While shared mental models drive the demand side dynamics of collective action, the mobilization of the 'critical mass' of human and shared knowledge networks is a technological prerequisite for the supply side dynamics of generalized collective action. In other words, the organizational nucleus for any collective action is constituted if a group of well-endowed actors invest their resources upfront to form a critical mass of human and shared knowledge network. This network then draws other actors into the organization and increases its power and productivity.

Once the 'critical mass' of resources is invested, other groups of actors may consider it to be in their 'enlightened self-interest' to participate in the collective action and produce the 'common good'. The first-mentioned group of actors is called 'focal actors' and the second-mentioned group is called 'potential actors'.

3.4.2 Potential Actors and Focal Actors

Potential actors are common people in the community who are potential beneficiaries of collective action. They are ready to exercise collective choice if it is consistent with their enlightened self-interest. They are the groups of individuals who are called non-activists in Oliver and Marwell (1992). They provide the essential, non-substitutable labor input without which no organized collective action can bear fruit or sustain itself.

Focal actors[9] are those individuals who have a higher level of motivation for collective action, being richly endowed with shared knowledge or 'shared mental model' (Denzau and North, 1994) about the need for such an action. They provide the knowledge-capital input in the production function and increase productivity of the efforts of potential actors. Driven by altruism, a sense of justice, duty, commitment, and social responsibility, they strive to achieve their non-material well-being through collective action. By filling the *structural holes* in knowledge and human networks, they mobilize potential actors and motivate them to bring about the desired collective good for common benefit.

Focal actors are knowledgeable but need not be idealistic. They perceive critical insights and comprehend ideas, facts and causal connections that other people cannot perceive or comprehend. Endowed with their knowledge, foresight, credibility and trustworthiness, they disseminate their shared mental models among people. They manage to change the structure of situations and frames of reference for perceiving the incentives or payoff structure that together lock people into non-cooperative PD type behavior. Driven by their preference for non-material well-being, they work single-mindedly to resolve collective action problems and bring about the collective good.

By accessing focal actors' human and knowledge networks, common individuals can develop preference for enlightened self-interest and turn into 'potential actors' who get motivated to exercise collective choice of cooperative behavior and participate in the collective action. Credibility/trustworthiness/charisma enables focal actors to evoke potential actors' trust and faith that together they can bring about an alternative set of behavioral norms that can replace the existing inferior ones. Such focal actors are able to draw trustworthy people around them, disseminate shared knowledge and form the initial nucleus of trustworthy actors for initiating collective action. This 'nucleus of trust' (Ostrom, 2000) draws different groups of potential actors (economic interest groups or social groups based on caste, religion or gender) into a wider network of potential actors.

Institutions such as the press, electronic and social media, judiciary, audit agencies and civil societies can also be drawn into potential actors' networks if they access 'shared knowledge' or 'shared mental models' and choose to disseminate them among the general public. Some of these institutions can also turn into focal actors themselves. By filling 'structural holes'

[9]Focal actor(s) may or may not be leaders or entrepreneurs. Unlike leaders or entrepreneurs, focal actors may not be overly concerned with any power equation or authority status or cost-benefit calculus. They are focused only on investing their knowledge, time and material resources with the intent to bring about the desired collective good.

(Burt, 1992) in the nascent networks of these different groups of potential actors, focal actors connect non-redundant contacts in the human network, as a result of which the power of the network to generate collective benefit is enhanced. The diffusion of 'shared knowledge' and ideologies or 'shared mental models' through the human and knowledge network of focal and potential actors augments the network power that drives collective action.

3.4.3 Shared Knowledge and its Accessibility

The endogenous development of preferences and behavioral choices depends on two state variables: i) the stock of shared knowledge or 'shared mental models' (Denzau and North, 1994; Mantzavinos *et al.* 2004); and ii) the stock of people (human capital) who can access such shared knowledge or mental models. Two flow variables correspondingly emanate from these two stock variables as focal actors access the knowledge stock, internalize it, refine or develop it further: i) new or incremental knowledge capital: ii) incremental accessibility to the stock of shared knowledge. New knowledge-capital and higher accessibility to shared knowledge convert more and more common people into potential actors, increasing the power of knowledge and human networks for collective action.

Shared mental models have both static and dynamic dimensions (Mantzavinos *et al.* 2004). Static dimensions are formed as 'individuals in a given socio-cultural environment continually communicate with other individuals while trying to solve their problems' (Mantzavinos *et al.* 2004, p.76). Static dimensions of shared knowledge are context specific and therefore critically important in the operational management of any particular collective action. On the other hand, the dynamic component of the shared mental model evolves across generations. The knowledge developed in the process of working with shared mental models over a period of time gives rise to the dynamic 'shared mental model' or shared knowledge that continually grows through learning across generations.

All societies at all points in time have had some perceptive individuals who could readily access and internalize stocks of shared knowledge, both static and dynamic, and develop new knowledge capital including insight or wisdom, which causes endogenous development of their non-material preferences. Driven by their non-material preferences, these individuals turn into focal actors and voluntarily interact with common people in order to help them access the shared knowledge about the potential benefits of a particular collective action in a given context. In this process the 'nucleus of trust' is constituted for the organization of collective action.

The institutionalization of this process of interaction and dissemination of shared knowledge/mental models through the formation of dense and centralized networks (Marwell *et al.* 2008) is the key to collective action. Since there is little incentive for the formal educational system to impart shared knowledge or mental models that have no market value, common people have no means to access such knowledge other than through focal actors' networks. Focal actors on their part willingly invest their resources and develop knowledge and human networks by building bridges with networks of groups of potential actors including the press, media, intellectuals, civil societies and other influencers. Through these networks, focal actors interact continually with potential actors and develop the static dimension of a shared mental model in a particular context.

Such a shared mental model gives 'a common interpretation of reality' (Mantzavinos *et al.* 2004, p. 76) as regards the contextual benefit of the collective action for common people. This common interpretation of reality fills the structural holes in the combined networks, as a result of which the diffusion of shared knowledge or mental models is institutionalized at little or no social cost. Such institutionalized diffusion of shared knowledge helps in the development of endogenous preferences for collective action among a segment of common population who accesses the shared knowledge through focal actors' knowledge and human networks and develops enlightened self-interest as they begin to comprehend the potential benefits of collective action and turn into potential actors.

If an individual has enough knowledge endowment to internalize the shared knowledge fully, he may also develop a strong endogenous preference for non-material well-being and become a focal actor in sustaining the collective action over time. Focal actors thus generate positive external effects of shared knowledge, which are internalized by newer actors as they fuel the growth of the focal actors' network and sustain collective action. A focal actor can also generate 'herd externality' for groups of potential actors connected in his network. These groups become influenced by the focal actor's thoughts and ideas, provided that the focal actor develops enough 'legitimacy capital' (Ray, 2012) in the course of the focal actor's continual interactions with them. Through these interactions, the focal actor develops the 'static dimensions' of 'shared mental models' (Mantzavinos *et al.* 2004).

In summary, shared knowledge and its accessibility through focal and potential actors' networks are critical factors in the production of collective action. Shared knowledge can in principle be acquired in three ways: the educational system, mass media and focal actors' knowledge networks. Educational institutions and mass media are often captured by vested interests that usually benefit from the *status quo*. Moreover, educational institutions may not have any incentive to disseminate such knowledge, as it has little or no market value and is not skill enhancing in any commercial sense. Therefore, shared knowledge and its diffusion through focal and potential actors' networks are the critical factors that initiate and sustain a collective action.

3.4.4 Prevailing Objective Conditions in the Society

A particular set of initial objective conditions in the society may facilitate collective action while another set may impose constraints. Favorable conditions include the following: prevailing social capital resource endowment of the society such as the number of associations, organizations and other social platforms for communication and social interactions; the degree of state and market failures that trigger the need for collective action—the higher the degree of failure the higher is the potential benefit of collective action; the level of democratization of society allowing exercise of the endogenously developed choice even as it may hurt vested interests—this is reflected in the freedom of institutions such as the press, media, judiciary, educational institutions and civil societies to advocate for change rather than the *status quo*; the existing stock of shared knowledge and its accessibility among common people; existing networks of focal actors with preference for non-material well-being; and opportunities to develop accessibility or capability to access the knowledge and human networks of focal actors.

Another class of objective conditions in the society and community may inhibit collective action: rigid caste division, gender division, social and economic inequality, and powerful vested interests that benefit from the *status quo*. Most importantly, some well entrenched social norms (such as corruption and the feudal system of big landlords providing financial assistance to agricultural laborers in times of need), formal institutions (such as subsidized water and electricity rates, electoral funding by the rich and powerful), informal institutions (such as unlimited extraction of groundwater below a landed property as an easement right flowing from the property right on the land) and policy enforcement regimes (such as violating with impunity water rights and arrears of water fees) prevent the poor and disenfranchised from organizing collective action, as they fail to grasp the complexities in the structure of situations and decipher cost effective ways of solving their problems.

The vast majority of the poor and the subaltern do not have access to the knowledge and resource endowment that could untangle the complicated problem space and identify the injustice and gross legitimacy deficits in the structure of the prevailing social system. Sometimes when they do get some access to such knowledge and other resource endowments, they may be so engaged in meeting their bare survival needs that they do not have enough time to internalize the knowledge and self-organize collective action. Even if they can somehow put together an organizational platform with the help of financial and knowledge resources of focal actors, they may not be able to participate effectively in the structuration and legitimation process dynamics that can alter the prevailing structure of situations. As a result they fail to put in place a legitimate institutional infrastructure suitable in the given context that can sustainably manage the generation as well as distribution of collective benefit.

A higher level of material prosperity of a society facilitates access and dissemination of shared knowledge in many ways: i) such a society at an aggregate level (the state, civil societies, independent press, electronic media, web-based social media, or independent judiciary) can afford to capitalize the shared knowledge as a public good, and disseminate it widely by developing appropriate institutional mechanisms to increase its accessibility among the general population; ii) individuals in such a society can afford to invest resources as focal actors and build the requisite knowledge and human networks; and iii) more people in such a society can afford to become potential actors and pursue their enlightened self-interest since there is no immediate threat to their basic needs of food and security.

However, relative material prosperity is not a sufficient condition for organizing collective action. More affluent societies may in fact be locked into an economic and political system that benefits the rich and middle-class population while the vast majority of the poor remain excluded as they cannot get access to the networks. Being impoverished, disenfranchised and powerless, they are locked in a perpetual low-level equilibrium from which only collective action through self-organization can pull them out. And in order to self-organize for collective action, they need to access the shared knowledge and mental models of the gross legitimacy deficit in the prevailing structure of situations and comprehend the potential benefit from their collective action.

Three structural inadequacies in the objective conditions in the society give rise to collective action problems: i) inadequate access to shared knowledge; ii) inadequate supply of human network of credible focal actors; and iii) inadequate ability of potential actors to adapt the knowledge in a given context. These lie at the heart of collective action problems. The problem of perverse incentives such as those in the PD game, the Olsonian model of free riding behavior in the provision of common good, and the Hardinian tragedy of the individually 'rational' overconsumption of common resources are only symptoms of these three more fundamental structural problems. All collective actions must address these problems and offer structural solutions in order to sustain such actions.

3.4.5 Sustaining Collective Action: Role of Structuration and Legitimation Dynamics

A collective action takes place at two levels: i) the constitutional level, and ii) the governance or operational level. At the constitutional level, potential actors self-organize themselves and the core nucleus of the human and knowledge network of the focal actor and potential actors is constituted through structuration process. At the operational level, managerial instrumentalities are deployed by the potential actors to fructify organizational goals, namely, producing the desired collective benefit and sharing the benefit in a legitimate manner.

Sustaining any collective action necessitates sustaining three interconnected process dynamics at these two levels: i) the 'structuration' process[10] at the constitutional level, including the continual development of focal actors' knowledge and human network and the internalization of knowledge and adaptation of shared knowledge in the prevailing societal context; ii) the legitimation process at the operation level that continually evaluates the legitimacy of the operational actions and their outcomes in terms of their congruence with the needs of potential actors and the expectations of other stakeholders such as focal actors and state; and iii) reflexive two-way feedback between the structuration process at the constitutional level and the legitimation process at the operational level that may either reinforce or weaken the structuration and legitimation process dynamics.

Through these process dynamics, focal actors develop new knowledge and static mental models (through the structuration and legitimation process) and disseminate dynamic as well as static mental model or shared knowledge among potential actors. The potential actors then apply the knowledge and mental models with such necessary adaptations as may be considered desirable, proper and appropriate in the given context (the legitimation process). Most collective actions fail at the governance level on account of the absence of this mutually self-reinforcing legitimation and structuration dynamics, and this absence happens due to ineffective coordination between these two dynamic processes at two different levels, the constitutional and operational levels.

There are four prerequisites to effective coordination of the ongoing process dynamics at these two levels: i) collective goods produced at the operation level must be tangible and the benefits must be distributed in such a way that the legitimate needs of concerned potential actors and the expectation of focal actors are met. This necessitates effective management of the legitimation process: ii) development and refinement of shared knowledge and common people's accessibility to and understanding of the knowledge must continue to grow; iii) focal actors must remain focused on their role, namely, the refinement, development and dissemination of shared knowledge about the collective action, and the development of its accessibility. These structuration dynamics at the constitutional level are required to be effectively managed by focal actors so as to reaffirm the validity, credibility and trustworthiness of their actions in the eyes of the potential actors; and iv) objective social conditions remain conducive to social learning and the formation of endogenous preference structure.

It was because of the absence of these factors that socialist revolutions in the USSR and China failed to sustain themselves as collective action of the working class people. Sustaining collective action against poverty, environmental degradation and corruption has also been difficult in several countries because of the absence of these factors.

---

[10] Giddens (1984) offers one way of conceptualizing this process dynamics, while Grewal (2008) offers a somewhat different way.

To sum up the general theory of organization of collective action, four separate yet interdependent pillars constitute its foundations: i) endogenous individual preferences and choice; ii) human and knowledge network of focal actors and potential actors; iii) shared knowledge about collective action and its accessibility; and iv) prevailing objective conditions including the stock of human and knowledge networks and presence of democratic institutions that empower common people. The key agency in this organization is focal actor, who, being endowed with shared knowledge and a strong preference for non-material well-being, makes upfront investment and initiates the building of a network of credible and trustworthy actors, what Elinor Ostrom referred to as a 'nucleus of trust'. Collective action is self-organized through this human and knowledge network.

While focal actors are rare individuals gifted with high knowledge endowment and an idiosyncratic need for enhancing their non-material well-being, there are more common individuals, potential actors, who can transform the free-riding behaviour that maximized their short-term narrow self-interests into a collective behavioral choice that can maximize the collective interest of the society and thereby their enlightened self-interest. More potential actors develop an endogenous preference for cooperation as they access the shared knowledge and human network and begin to believe that they have much to gain by participating in the collective action. They represent common beneficiaries without whose continued participation no collective action can be sustained. Finally, no collective action is sustainable without removing some constraining objective conditions such as entrenched institutions that impose prohibitive costs on self-organization. Removal of such well entrenched institutions itself requires collective action at the structural level, aided by the structuration dynamics at the constitutional level and legitimation dynamics at the operational level.

3.5 APPLYING THE GENERAL THEORY FOR COLLECTIVE WATER RESOURCE MANAGEMENT: THEORETICAL PROPOSITIONS FOR EMPIRICAL VALIDATION

In the context of participative water resource management, focal actors are either local social activists and/or voluntary organizations, national and transnational NPOs and international organizations. Potential actors are actual water users who cooperate in participative management of water resources. Institutional designs and principles behind those designs such as Elinor Ostrom's design principles are examples of shared knowledge and mental models. The relevant objective conditions in the society that facilitate or constrain participative water resource management are the regulatory framework provided by the state and the power equations between potential actors' network and vested interests.

Typically, NPOs/international organizations make initial investments in terms of sharing their knowledge and experiences about efficient and effective participative water management practices and helping water users in designing the operational rules and management practices in collective water management projects. They constitute the core human and knowledge network accessing which water users, who were inclined to over-consume or waste water, begin to change their behavior and start cooperating in efforts for conserving the resource through its effective and efficient participative management by the network of focal and potential actors. Knowledge acquired through social learning changes their preference structure and choice for collective action as they begin to be aware of the adverse social implications of the ongoing water abuse. This change in behavior is due not to the threat of statutory punishment but to an increased level of collective knowledge and understanding about the water stress and the future disastrous consequences of inefficient and wasteful use of water.

The NPOs/international organizations/social activists also help the community of water users to frame the rules for participative water management that are effective in their local context. They also help the community to connect with state agencies by filling the information gaps (structural holes) between them, thereby increasing the network power of the community of water users. Sometimes committed state agencies may also become part of the focal actors' network

Finally, the NPOs/international organizations participate effectively in both the structuration and legitimation processes, including dispute resolution and fair and equitable water consumption. In developing countries, they also finance the initial capital expenses for capacity development without which poor water users cannot effectively participate in the social learning processes for management of water resources. All these discussions lead to Proposition 1

Proposition 1

NPOs, international agencies, committed state authorities and dedicated social workers constitute the core knowledge and human network or the 'nucleus of trust' necessary for organizing collective action for efficient and effective participative water resource management.

Organizing collective action becomes challenging at the constitutional level itself in certain complex situations fraught with uncertainty about the distribution of potential future benefits of a particular collective action. Some illustrative examples are integrated water resource management (IWRM) programs, linking of river basins, and management of trans-boundary river basins. In such cases, two structural drivers of collective action become critically important: i) the accessibility of potential actors (in the case of IWRM within a state/country, potential actors are users of different water resources such as groundwater aquifers, canals, water reservoirs such as tanks and ponds, and watersheds) to the shared knowledge and human networks of focal actors (in the case of IWRM, focal actors are international organizations or their sponsored agencies, representative of local governments and expert technical bodies such as committees appointed by national governments); and ii) credibility of focal actors' knowledge and human networks.

According to the general theory, uncertainty about the net benefit of a collective action is mitigated by social learning through an effective dissemination of focal actors' knowledge about the structure of situations causing the uncertainty. For such an effective dissemination, potential participators must have an improved accessibility to focal actors' human and knowledge networks, which they must find comprehendible, workable and credible. This in turn requires capability development of potential actors, continual knowledge development on the part of focal actors, and most importantly the relevancy, suitability and legitimacy of the shared knowledge and mental model in governing the common water resources so as to address the need and expectations of all actors and resource providers.

As potential participators or actors find the knowledge content and human resources of NPOs/international organizations increasingly credible, more such actors voluntarily participate in the organized collective action. Greater participation of water users increases the level of expected tangible benefits, mitigates the uncertainty and strengthens the focal actor-potential actors combined network for organizing collective action. This leads to Proposition2

Proposition 2

Higher accessibility to and credibility of human and knowledge networks of NPOs, international organizations, dedicated social activists and committed state agencies are critically important for the constitution of an effective collective action in complex problem space such as IWRM, linking of river basins and trans-boundary river basins.

The demand for the organization of collective action at the stage of its constitution comes from its two drivers: i) the expected net benefit from collective action; and ii) the social norm of cooperative behavior which in turn comes from the preference structure of potential actors in the community that evolves endogenously from their social learning experiences.

As NGO and/or state agencies create a platform or network for exchange of knowledge and information about the water stress created by inefficient and wasteful water use, water users access the shared knowledge and change their behavioral choice endogenously to become cooperative and rule-compliant users. Water users self-organize and put into practice the shared knowledge and mental models as internalized and adapted by them in order to make the model suitable for effective and efficient participative management of water resources in the local context.

The critical driver of the demand for organizing collective action at the constitutional level is water users' social learning and consequent behavioral change caused by the structural transformation of their preference through a structuration process that is reinforced by the legitimation process at the governance level. The structuration process is triggered by water users' belief and understanding that collective benefit can be successfully harnessed by applying focal actors' knowledge and human network. Shared knowledge/mental model, rather than the threat of punishment in formal institutions, and its internalization by potential actors, constitute the organization of the human and knowledge network for collective action.

As observed by Wade (1987) and Ostrom (2000), an appropriate institutional arrangement for sanctions and its effective enforcement is, however, an important objective condition that signals to rule-compliant water users that a legitimate institutional structure is in place to protect their interests and insure that they do not become 'suckers'. Proposition 3 emerges from these discussions.

Proposition 3

Collective action is sustained if the rule-compliant behavior of water users evolves as a social norm from the structuration dynamic that is driven by i) their access to the shared knowledge and shared mental model or belief about the net tangible benefits of collective action, and ii) feedback of the existence of legitimate governance process through monitoring of the net benefit of collective action and the distribution of the benefit.

Proposition 3 may not, however, necessarily apply in the case of presence of powerful vested interests (such as industries and the large landowning class) because they may work against the legitimate governance process. This brings into play the prevailing power equation in the socio-political space, an equation between the class interests of capitalists/feudal lords and the combined network power of common water users and focal actors. In the presence of truly democratic institutions and visible adverse impacts of the large-scale appropriation of common water resources by these vested interests, the power equation may tilt in favor of common water users.

Under such objective conditions, public pressure can be brought to bear upon vested interests, either through the state or through direct collective action in the form of a people's movement for the legitimate use of common water resources by different interest groups. Such a collective action aimed at breaking the stranglehold of vested interests is facilitated if common people have pre-existing social platforms, media networks and a network of focal actors, including committed voluntary organizations, public intellectuals and dedicated social activists. These discussions lead to Propositions 4 and 5.

Proposition 4

Collective action against vested interests of large appropriators of common water resources such as industries and large agricultural firms is constituted only under truly democratic conditions in the society, which makes the network power of legitimate water users and committed focal actors strong enough to challenge the power of vested interests.

Proposition 5

Institutional arrangements for the punishment of rule violators and its effective enforcement increase the legitimacy of the organized collective action and such legitimacy is critically important for continued participation of potential actors in common water resource management at the operation and management levels.

Sustaining collective action necessitates the continual flow of resources from water users in terms of their efforts towards participative management of water resources as well as knowledge and financial resources of NGOs, state agencies and dedicated social activists, which happens only if the action and its outcome appear legitimate from resource providers' perspectives. Legitimacy of a collective action means congruence of operational and management practices, their beneficial outcomes and the distribution of benefits with social norms, values, beliefs, needs and expectations of the community of water users as well as focal actors. The flow of tangible benefits of water use from collective water resource management is therefore a necessary condition, while just and fair distribution of its fruits among all water users is a sufficient condition for sustaining collective management of common water resources. Propositions 6 and 7 can be deduced from these discussions

Proposition 6

Legitimacy evaluation of the governance process and outcome of water management practices is necessary for minimizing conflicts and sustaining effective collective action in the management of water resources.

Proposition 7

Collective action in the management of water resources can be sustained if and only if there exist self-reinforcing feedback dynamics between the structuration dynamics (the process of development of new knowledge and mental models and sharing them with common people through social learning programs) at the constitutional level and the legitimation dynamic (the process of legitimacy evaluation) at the governance or operational level.

3.6 CONCLUDING REMARKS

Although there are some prominent success stories of collective action in water resources in both developing and developed countries, efforts to mobilize and sustain collective action in managing water resources have failed more often than they have succeeded. Most of these efforts were directed towards institutional arrangements based on the design principles enunciated in Elinor Ostrom's works on common pool resources. Some World Bank funded efforts in the 1990s also focused on Robert Wade's set of conditions for the successful organization of collective action in the management of common property resources. But success stories of participative management of common water resources were found to be not replicable even though Ostrom's set of design principles was applied in such settings where the set of favorable conditions as observed by Robert Wade appeared to exist.

One of the weaknesses of the extant literature on participative management of common property resources is its overemphasis on the efficacy of institutional design in changing the structure of situations that generate perverse incentives for non-cooperative behavior on the part of 'rational' appropriators. But it does not provide the micro-foundations for the supply of such institutional design. Who will design such institutions and why? What is the process dynamics through which appropriate design principles suitable in a particular context may emerge?

The extant literature also does not throw any light on the micro-foundations of the demand for collective action. Why and how will appropriators of common property resources change their preference for self-centered behavior and start thinking in collective terms? If the fear of punishment does the trick, then collective action is not really voluntary and therefore not sustainable. The literature neglects structural dynamics, such as social learning that can alter the preference structure of individuals and motivate them to cooperate in collective efforts for restricted aggregated use and regeneration of the common resource rather than free ride on others' efforts.

The dominant strand of thought in the theoretical literature on collective action focuses on solving three interrelated collective action problems captured in Mancur Olson's seminal book *The logic of Collective Action*, the PD game, and Garett Hardin's 1968 classic "The Tragedy of Commons". Research issues were framed accordingly for theoretical analysis and empirical investigation. But these research agenda and their findings do not fit in situations where these three models do not really capture the real problems on ground. In fact, as Oliver and Marwell's work and other analyses have suggested, free riding may not be a real problem in most settings.

Similarly, the analytical framework in 'Hardinian tragedy' is based on the assumption that individual appropriators are shortsighted and utterly selfish. This is a narrow and one-sided view of humanity. Such a view may not be realistic in most community settings, particularly in Asia. Thus the theoretical propositions and frames of research agenda that emerge from such a narrow perspective of collective action problems and the findings from such lines of research turn out to be irrelevant in most real-life situations.

There has also been no attempt in the literature on collective action to explain why and how some organizations for collective action survive in the long run while others die quickly. The notion of organizational legitimacy can throw useful light on the sustainability of the organizations for collective action. Clearly, there are many gaps in the current body of literature on collective action theory. As a result of these deficiencies, research issues are not being framed appropriately for empirical investigation. There is a need to address these inadequacies in the analytical frameworks of collective action and reframe research issues for empirical investigation. This calls for a new theoretical framework of generalized collective action that is valid in diverse settings for different classes of collective action and for extracting a new set of deductive propositions from the general framework as applied to a specific class of collective action. This work is a step in that direction.

3.7 REFERENCES

Bardhan P. (1993). Symposium on management of local commons. *Journal of Economic Perspectives*, **7**(4), 87–92.

Basu K. (2011). *Beyond The Invisible Hand*. Princeton University Press, Princeton.

Burt R. S. (1992). *Structural Holes: The social Structure of Competition*. Harvard University Press, Cambridge, MA.

Davis D. D. and Holt C. A. (1993). *Experimental Economics*. Princeton University Press, Princeton NJ.

Denzau A. T. and North D. C. (1994). Shared mental models: ideologies and institutions. *Kyklos*, **47**(1), 3–31.

Dombrowsky I. (2007). *Conflict, Cooperation and Institutions in International Water Management: An Economic Analysis*. Edward Elgar, Northampton, MA, USA.

Frohlich N., Oppenheimmer J. A. and Young O. (1971). *Political Leadership and Collective Goods*. Princeton University Press, Princeton, NJ.

Giddens A. (1984). *The constitution of Society: Introduction to the Theory of Structuration*. University of California Press, Berkley.

Gleick P. H. (1993). Water and conflict: fresh water resources and international security. *International Security*, **18**(1), 79–112

Grewal D. S. (2008). *Network Power: The Social Dynamics of Globalization*. Yale University Press, New Haven and London.

Guth W. and Kliemt H. (1998). The indirect evolutionary approach: bridging the gap between rationality and adaptation. *Rationality and Society*, **10**(3), 377–399.

Hardin G. (1968). The tragedy of the commons. *Science*, **162**(3598), 1243–1248.

Ledyard J. (1995). Public goods: a survey of experimental research. In: *Handbook of Experimental Economics*, J. Kagel and A. Roth (eds.), Princeton University Press, New Jersey.

Mantzavinos C., North D. C. and Shariq S. (2004). Learning, institutions, and economic performance. *Perspectives in Politics*, **2**(1), 75–84.

Marwell G., Oliver P. E. and Prahl R. (1988). A theory of critical mass III, *American Journal of Sociology*, **94**(3), 502–534.

Offerman T. (1997). *Beliefs and Decision Rules in Public Goods Games: Theory and Experiments*. Kluwer Academic Publishers, Dordrecht, the Netherlands.

Oliver P. E. and Marwell G. (1988). A theory of critical mass II. *American Sociological Review*, **53**(1), 1–8.

Oliver P. E. and Marwell G. (1992). Mobilizing technologies for collective action. In. *Frontiers in Social Movement Theory*, A. D. Morris and C. M. Mueller (eds.), Yale University Press, New Haven, 251–72.

Oliver P. E., Marwell G. and Teixeira R. (1985). A theory of critical mass I: interdependence, group Heterogeneity, and the production of collective action. *American Journal of Sociology*, **91**(3), 522–556.

Olson M. (1965). *The logic of Collective Action: Public Goods and the Theory of Groups*. Harvard University Press, Cambridge, MA.

Ostrom E. (2000). Collective action and evolution of social norms. *The Journal of Economic Perspectives*, **14**(3), 137–158.

Ostrom E. (1990). *Governing the Commons: the Evolution of Institutions for Collective Action*. Cambridge University Press, New York.

Ostrom E. and T.K. Ahn. (2003). Introduction. In. *Foundations of Social Capital*, E. Ostrom and T.K. Ahn (eds.), Edward Elgar, Cheltenham, UK Northampton, MA, USA. pp. xi–xxxix.

Ray G. (2012). Legitimacy capital and development: a theoretical inquiry. *Paper presented at The Ninth Workshop on Social Capital and Development Trends in the Swedish and Japanese Countryside* held in Österlen, Sweden on 24–25 May, 2012.

Ray G. (2013). Foundations of Collective action: towards a general theory. In: Social capital and rural development in a knowledge society, H. Westlund and K. Kobayashi (eds.), Edward Elgar, London.

Sen A. K. (2006). *Identity and Violence: The Illusion of Destiny*. W.W. Norton, New York.

Starr J. R. (1991). Water wars. *Foreign Policy*, **82**(spring), 17–36.

Wade R. (1982). The system of administrative and political corruption. *Journal of Development Studies*, **18**(3), 287–328.

Wade R. (1987). The management of common property resources: collective action as an alternative to privatization or state regulation. *Cambridge Journal of Economics*, **11**, 95–106.

4 Participatory approach to community based water supply system

Kenshiro OGI[1], *Ismu Rini Dwi ARI*[2], *Hayeong JEONG*[3], *Kakuya MATSUSHIMA*[4], *Kiyoshi KOBAYASHI*[5]

[1] Department of Urban Management, Graduate School of Engineering, Kyoto University. E-mail: kenshiro.ogi@gmail.com

[2] Department of Regional and Urban Planning, Faculty of Engineering, University of Brawijaya. E-mail: is_2mu@yahoo.com; dwiari@ub.ac.id

[3] Department of Urban Management, Graduate School of Engineering, Kyoto University. E-mail: jeong.hayeong.2z@kyoto-u.ac.jp

[4] Department of Urban Management, Graduate School of Engineering, Kyoto University. E-mail: matsushima.kakuya.7u@kyoto-u.ac.jp

[5] Department of Urban Management, Graduate School of Engineering, Kyoto University. E-mail: kobayashi.kiyoshi.6n@kyoto-u.ac.jp

Abstract

This is an empirical research with main aim to investigate interdependent preference in a choice of access to water supply system in the community with respect to Indonesian rural areas covering Toyomarto and Candi Renggo village in Singosari district, Malang regency. The sample of respondent is classified into two types based on their choice to water supply system. First is the respondents who participate in CBWM - community based water management (HIPPAM's member and get water from public hydrant), and second is the respondents who do not participate in community based water management or for those which obtain clean water from different access (PDAM, individual well and others). This research applies spatial probit model with posterior distribution from the Bayes theorem and express the MCMC sampling method. From the estimation results, it might conclude that the length of living in the area and the price of monthly water usage have an important meaning for respondent to make a decision to join community based water supply system.

Keywords: participation approach, community managed water supply system, spatial probit model, Bayesian inference, MCMC

4.1 INTRODUCTION

Regarding Network of Asian River Basin Organizations (NARBO, 2007) reports, about 600 million people in the Asia-Pacific region do not have access to clean water. Rapid growth of economic and population in Asian developing countries sort of China, India, and Indonesia generate an explosive growth in water demand. Therefore, it is most pressing challenge to search for a safe, secure and reliable water supply measure.

In addition, every country has been searching for water supply systems for securing water accessibility for citizen through infrastructure improvement and expanding its services. Nevertheless, it is absolutely not easy to distribute a limited alternative water resource properly and objectively under the budget restrain of government. In order to address the problem, it has the development of technology but what's more enhancement of collaborative governance for water management is essential.

In the case of Indonesia, among a total of 231.6 million inhabitants, only around 49.7% have access to water, 10% of which is obtained through community-based water management (Statistic Centre Board, 2005). It might define as an interesting example of collective action. Collective action, which may be defined as action on the part of one or more people striving to achieve the objectives or satisfy the common interest of the group, implies devising frameworks that limit the pursuit of individual self-interest and sustain the benefit shared by the group. In order to support and sustain the participatory approach to community-based water supply systems, it is necessary to clarify mechanisms and to invent institutional systems to support collaboration activities. However, there has been insufficient investigation of the participatory approach to community-based water supply systems. It is urgently necessary to achieve a better understanding of local community structure and networks in order to realize suitable water policy and institutional restructuring.

Therefore this study questions the following: why people are involved in establishing a community-based water supply system in order to access water; whether their choice to join the community-based water supply system is cooperative or not; and why people do or do not engage in a kind of collective action. It is important to investigate the mechanism of spontaneous collaboration to access water. Using a field survey on Indonesia's water supply system, this study aims to investigate interdependent preference in the choice for clean water in the community.

The chapter is structured as follows. Section 2 illustrates the current situation of the respondent and access to water in the study area. Section 3 explains the model, which consists of a spatial probit model and the Markov Chain Monte Carlo (MCMC) as the estimation method. Section 4 describes the results and discusses empirical applications. Finally, the paper is concluded with section 5.

4.2 EMPIRICAL RESEARCH

To illustrate the model in an applied setting, this research used data from a field survey which was conducted in 2008. Through systematic sampling, 500 households in Toyomarto and Candi Renggo village, Singosari district, Malang regency, East Java Province Indonesia, were selected as respondents for the study (Figure 4.1). Data were collected employing a face-to-face interview method that was effectively completed within 10 days by 20 surveyors with interview schedules from 07:00 a.m. to 09:00 p.m., depending on the respondents' readiness.

The restraint of the study is that the respondents were selected from a total of 24,388 inhabitants in the two villages. Hence the results and the substantial assessments replicate the essential characteristics of the contacted people. Furthermore, in order to optimize the representativeness of the inhabitants, the respondents selected for the study were the husband, the wife or the head of the family chosen so as to represent typical important inhabitants.

Referring to the Instruments of the Social Capital Assessment Tools developed by Chiristiaan *et al.* (2004) and combining it with preliminary survey information, it developed a household questionnaire survey.The questionnaire survey covers questions that concern (i) demographics data of the respondent (gender, age, family size, occupation, length of stay, income, and education), (ii) access to water and water usage, and (iii) community network through their participation in community group andconcern for community welfare.

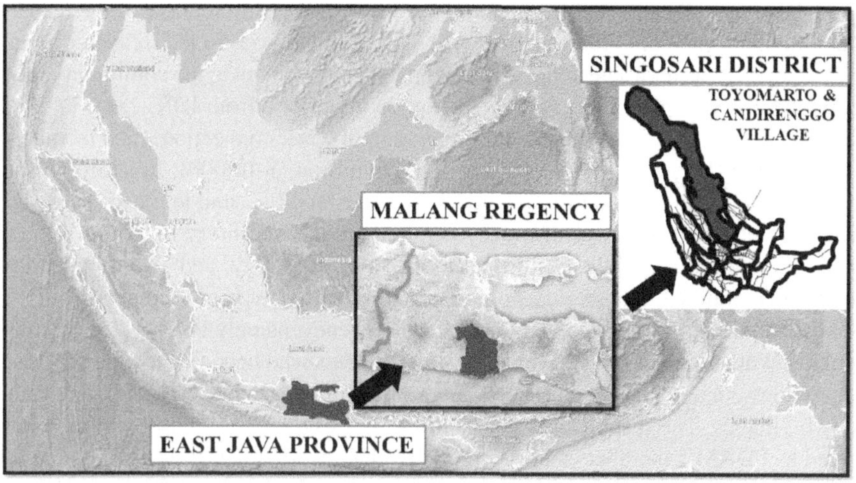

Figure 4.1 Location of the research area. (*Source*: Field Survey, 2009).

4.2.1 Access to Water

In general, there are two types of fresh water resources utilized by the respondents in the study area, namely surface water and shallow ground water.

Firstly, surface water consists of natural spring and river. There are three natural springs-the Candi Sumber Awan Spring, and Banyon Spring in Toyomarto village, and Kendedes Pond in Candi Renggo village. The other surface water comes from two rivers named Sumber Awan River and Petung River flowing away through the area of study. Secondly, shallow ground water is utilized by individual household who builds their individual well to get fresh water for domestic purposes. Based on categorization of goods, both types of fresh water resources could be categorized as common-pool resources (CPRs) which share a pair of common characteristics (i) their consumption is rival, but (ii) it is non-excludable (Wydick, 2008). Rivalry means that consumption by one person/group/ institution precludes consumption by another. In this case, water in the spring or river or shallow ground water that utilized by A person is one that B person can't consume it. Excludability means that it is easy to control access to a resource, and to exclude others from consuming it. A person can't exclude B person from using water in the spring, river, as well as shallow ground water to fulfill their domestic purposes.

A more detailed explanation follows of the five types of access to water in the research area. First, water services are supplied by groups engaged in community-based water management, namely HIPPAM (Resident Association of Drinking Water Users). HIPPAM serves the 42% of respondents who become members of certain HIPPAM groups. These community water providers utilize the three natural wellsprings to supply their members. One HIPPAM group encompasses around 30 to 300 households living in a certain neighborhood association through direct home water connection. This community group can be categorized as club goods consisting of a combination of non-rivalry and excludable. Non-rivalry here means that water consumption by one member permits water consumption by another. Excludability means that only the member has access to water from HIPPAM and excludes others from consuming it. The structural organization of HIPPAM encompasses 1 leader, 1 treasurer and 2 technician staff members annually selected from among all members. The main task of the leader is to maintain the management system of HIPPAM. The member goes to the treasurer to pay the monthly charge for water usage at the price set up under agreement at the community meeting. Water pricing per group employs various fixed monthly prices, from IDR. 2,000 (¥ 20) up to IDR. 15,000 (¥150). For the first physical installation of water connection, each member must pay from around IDR. 500,000 (¥5,000) to IDR. 1 million (¥ 10,000). In addition, specific tasks for technicians are to monitor and maintain water facilities, and to try to solve physical problems if they occur. Finally, all money compiled in the treasurery functions as a capital source to maintain the good the performance of the water facility, and also to pay a kind of service fee to committee members, whose jobs are basically a type of volunteer work.

Secondly, a drinking water company in Malang regency, namely PDAM (Local Company of Drinking Water), serves the 22% of survey respondents who become customers. This local water provider employs Candi Sumberawan wellspring to provide water for its customers in the two villages. The respondents who join PDAM's facility are those who are officially registered as PDAM's members and pay monthly charges for water usage. The price for the first physical installation of water connection is around IDR. 1,500,000 (¥15,000), and there is a fixed price for the first 10m^3 of water usage at IDR. 11,500 (¥115). Then, for subsequent water usage, the price per cubic meter for domestic purpose is IDR. 1,500 (¥15). The group of respondents who join PDAM are the group most likely to obtain fresh water access from PDAM as the local water provider. In this sense, water from PDAM could be categorized as club goods that are non-rival among PDAM members but which others are prohibitedfrom consuming.

Thirdly, there is access to clean water through private or individual wells. Residents build their own wells on their land property to obtain drinking water. This group accounts for 23% of the respondents, and this water can be categorized as pure private goods. Such residents usually utilize shallow ground water at a depth of less than 20 meters, with some residents also developing water pipeline installation from the well to the house (e.g., for the bathroom and kitchen), using electric pumps. Meanwhile, others have to draw water from the well manually, using plastic buckets which they carry to the kitchen or bathroom.

The fourth type of access to water is through public hydrants, which serve 12% of respondents. There are two types of public hydrant. First, several groups of HIPPAM, in addition to serving members, also build some public taps in the vicinity of residential areas, to serve residents without access to direct water pipeline connections to their houses. It is explicitly revealed that the purpose of the public tap is to serve non-members of community-based water management. The second type of public hydrant is a communal well where residents have to use plastic buckets to draw water from a facility built or provided by someone who likes to share his or her private goods with those who cannot afford their own wells. This circumstance shows the existence among groups in the community of bridging social capital, which Woolcock and Narayan (2000) refer to as the relationships we have with people who are not like us, i.e. relationships with people of different socio-economic status or ethnicity. In this sense, we could also discern how the community, through collective action, tries to get access to water.

The last type of access to water resources is used by respondents who still rely on rainwater or river water, such as from the Sumberawan River and the Glatik River, which flow through the two villages (1%). Please see Figure 4.2 for examples.

Figure 4.2 HIPPAM's water reservoir, communal hydrant, PDAM's water reservoir, individual well, river. (*Source*: Field Survey, 2009).

Table 4.1 describes the type of access to clean water of the respondents, 2 of whose data are excluded from the sample due to the inconsistency of their answers.

Table 4.1 Type of access to water.

Access to Water	Number of Respondent	Percentage
HIPPAM	210	42
PDAM	109	22
Individual Well	117	23
Public Hydrant	58	12
Others	4	1
Total	498	100

4.2.2 Demographic Data of the Respondents

There is a majority of male respondents (70%) as compared to females in the study. The average age of the respondents is 46.91 years, so that around 92% of respondents can be classified as of productive age (19–65 years). They have an average of 4 members in the household. Thirty percent of the respondents have occupations in the agriculture and manufacturing sectors, with typical livelihoods being food peddler, owner and worker in the small-scale sandal industry, farmer, breeder and field hand. Meanwhile, the majority of occupations could be categorized as belonging to the service sector, with various jobs such as private company employee (20%), labor industry worker (15%), and in the range of 1–6% merchant, driver, teacher, official, military/police, retiree, housewife, and even 4% of respondents with unemployment status.

Household income of a majority of the respondents is less than 1 million rupiah per month (64%), consisting of 25% of respondents with monthly income less than 500.000 rupiah and another 39% with monthly income between 500.000–1 million rupiah. This indicates that the residents are in the lower middle class in terms of socio-economic status, with the monthly minimum wage of Malang regency in 2009 having been determined at 945.500 rupiah.

Elementary schooling consists of six years of education, with three years in middle school and also in high school. According to education statistics (source: www.Nation Master.com), Indonesian adults have an average of five years of schooling. This implies that around 56% of the respondents in our study have an education level above the national average.

4.3 MODEL AND ESTIMATION METHOD

4.3.1 Model

This section focuses on a discrete-choice model for whether a household joins HIPPAM, conditional on that household's characteristics. We start by introducing the main assumptions in the model and notations that will be used for the rest of the paper. Let n be the number of individual households. Each household has two alternatives, labeled as 1 for joining HIPPAM and 0 for otherwise. For each household, we observe whether the household joins HIPPAM or not and model it as the realization of a random variable y_i. Economic theory suggests that the decision to join is primarily made to maximize the discounted value of future profits, so we assume that the choice of whether to join HIPPAM or not is the result of a household's decision to maximize their utility. An event will occur with a certain probability p if the utility derived from choosing that alternative is greater than the utility from the other alternative. Let z_i be the difference in utility between alternatives 1 and 0. The difference in utility is modeled as

$$z_i = x_i'\beta + \theta_i + \varepsilon_i \tag{4.1}$$

where $i = 1, \cdots n$, $x_i = (x_{ik}: k = 1, \cdots, K)'$ is a vector of observed household specific attributes, $\beta = (\beta_k: k = 1, \cdots, K)'$ is a vector of unobserved parameters to be estimated, θ_i is an unobserved random effect component, and ε_i is a stochastic error term with $\varepsilon_i \sim N(0,1)$. We do not observe z_i, but observe only the sign of z_i. We observe the household's choice y_i being equal to 1 or 0, depending on whether z_i has a positive sign indicating the higher utility from this alternative or a negative sign associated with the lower utility associated with this alternative. Therefore we observe

$$y_i = \begin{cases} 1 & if \quad z_i > 0 \\ 0 & if \quad z_i \leq 0 \end{cases} \tag{4.2}$$

The probability of choosing alternative 1 is given by

$$P_i = P\left(y_i = 1\right) = P(z_i > 0) \tag{4.3}$$

The distinction between this model and a standard probit model is the term θ_i. The unobserved component θ_i is constructed such that it allows for spatial interaction among households. This is obtained by specifying θ_i according to a spatial autoregressive structure (Smith and Lesage (2004), Lesage (2009)):

$$\theta_i = \rho \sum\nolimits_{j=1}^{n} w_{ij}\theta_j + u_i \tag{4.4}$$

with $u_i \sim N(0, \sigma^2)$, $W = (w_{ij}: i, j = 1, \cdots n)$ is a row standardized spatial weight matrix such that $\sum_{j=1}^{n} w_{ij} = 1$. ρ can be interpreted as the degree of spatial dependence across households. A positive (negative) value of ρ indicates a positive (negative) correlation among households. We can write equation (4.4) in matrix notation:

$$\theta = \rho W \theta + u, \tag{4.5}$$

where $u \sim N_n(0_n, \sigma^2 I_n)$ and I_n is the identity matrix. Letting $S = I_n - \rho W$, we can obtain a solution for θ using (4.5):

$$\theta = S^{-1} u. \tag{4.6}$$

It is worth noting that in our model, there is a network propagation effect captured in equation (4.5). Our model presents a simple test on the existence of the propagation effect. If ρ is significantly different from zero, then we conclude that there could be some spatial correlation beyond what is captured in the $x_i \beta$ term in the equation (4.1). From (4.6) we see that the distribution for θ_i is given by

$$\theta | (\rho, \sigma^2) \sim N_n(0_n, \sigma^2 (S'S)^{-1}). \tag{4.7}$$

The error term ε is assumed to be conditionally independent of the spatial unobserved component such that $\varepsilon | \theta \sim N_n(0_n, \sigma_\varepsilon^2 I_n)$ and we assume $\sigma_\varepsilon^2 = 1$. The full model in matrix notation is given by

$$z = X\beta + \theta + \varepsilon. \tag{4.8}$$

The likelihood function of this model is as follows:

$$L(y | \beta, \theta, \rho, \sigma^2) = \prod_{i=1}^{n} \Phi(x_i'\beta + (S^{-1})_i u)^{y_i} \left\{1 - \Phi(x_i'\beta + (S^{-1})_i u)\right\}^{1-y_i} \tag{4.9}$$

where Φ and $(S^{-1})_i$ respectively denote a cumulative distribution function of the standard normal and ith row of S^{-1}. But it is difficult to estimate this model by the maximum likelihood method since the likelihood function has a complicated form. Then, we use the Bayesian inference approach to estimate each parameter of the equation by using the Markov Chain Monte Carlo (MCMC) method (Gilks *et al*. 1996) that samples sequentially from the complete set of conditional posterior distributions for the parameters. The MCMC method provides a powerful tool for simulating complicated posterior distributions.

4.3.2 Bayesian Inference.

We estimated the above spatial probit model by using the Markov Chain Monte Carlo (MCMC) method. The first MCMC algorithm, popularly used in statistics and econometrics, was the Gibbs sampler, which arrives at the target distribution of unknown parameters by sequentially sampling from a set of conditional distributions of parameters. This is very useful since usually it is difficult to find an analytical result for posterior densities. The MCMC method provides a sample from the posterior density and we can use this sample to draw inferences about the parameters of interest. Under the mild regularity conditions satisfied in this application, these samples converge with the sample from the posterior distribution.

Most parameters can be sampled by using the Gibbs sampler (Gelfand and Smith, 1990); however, sampling the spatial parameter ρ is solely difficult since the conditional posterior distribution is not reducible to a standard distribution. Therefore, we apply a Metropolis-Hastings (MH) sampling method (Chib and Greenburg, 1995).

To derive the conditional posterior distributions, we use the Bayes theorem:

$$p\left(\beta,\theta,\rho,\sigma^2, z\middle|y\right)\propto L(y|\beta,\theta,\rho,\sigma^2,z)\cdot\pi(\beta,\theta,\rho,\sigma^2,z) \tag{4.10}$$

where $p(\cdot)$ represents posterior densities. The prior distributions of each parameter β,ρ,σ^2 are assumed independent; therefore, the posterior joint density $p(\beta, \theta, \rho, \sigma^2, z/y)$ is given up to a constant of proportionality by

$$p\left(\beta,\theta,\rho,\sigma^2, z\right)\propto L(y|z)\cdot\pi(z|\beta,\theta)\cdot\pi(\theta|\rho,\sigma^2)\cdot\pi(\beta)\cdot\pi(\rho)\cdot\pi(\sigma^2) \tag{4.11}$$

Using this relation, we obtain the appropriate conditional posterior distributions for each parameter in the model and examine MCMC sampling methods in the following section. Before we examine the Bayesian estimation we set each parameter's prior distributions as follows:

$$\begin{aligned}&\pi(\beta)\sim N_K(c,T),\ \pi(\sigma^2)\sim IG(\alpha,\nu),\ \pi(\rho)\sim U\left(\lambda_{min}^{-1},\lambda_{max}^{-1}\right)\\&\pi(\theta|\rho,\sigma^2)\sim N_n\left(0_n,\sigma^2\left(S'S\right)^{-1}\right),\ \pi(z|\beta,\theta)\sim N_n(X\beta+\theta,I_n)\end{aligned} \tag{4.12}$$

where β has normal conjugate prior distribution with means set to zero and covariance matrix set to $100I_K$, and σ^2 is assigned a conjugate inverted gamma prior with $\alpha = 25$ and $\nu = 3$. We employ a uniform prior distribution on ρ over a specified range. The parameter ρ must lie in the interval $\left[\lambda_{min}^{-1},\lambda_{max}^{-1}\right]$, where λ_{min} and λ_{max} denote the minimum and maximum eigenvalues of W, for the matrix $S = I_n - \rho W$ to be invertible. Introducing each prior distribution (4.12) into equation (4.11), we can derive the conditional posterior distributions for each parameter. In the next section, we examine the MCMC sampling method, using these conditional posterior distributions.

4.3.3 The Markov Chain Monte Carlo (MCMC) Sampler

The MCMC estimation scheme involves starting with arbitrary initial values for the parameters which we denote β^0, θ^0, $\sigma^{2(0)}$, ρ^0 and the latent variable z^0. We then sample sequentially from the following set of conditional distributions for the parameters in our model.

a) Calculate $p(\beta|\rho^0, \theta^0, \sigma^{2(0)}, z^0, y)$ using each initial parameter. We carry out a multivariate random draw to determine β^1.

$$\beta|(\theta, \rho, \sigma^2, z, y) \sim N_K(A^{-1}b, A^{-1}) \tag{4.13}$$

where $A = X'X + T^{-1}, b = X'(z - \theta) + T^{-1}c$.

b) Calculate $p(\theta|\beta^1, \rho^0, \sigma^{2(0)}, z^0, y)$. We carry out a multivariate random draw to determine θ^1.

$$\theta|(\beta, \rho, \sigma^2, z, y) \sim N_n(A_0^{-1}b_0, A_0^{-1}) \tag{4.14}$$

where $A_0 = \sigma^{-2} S'S + I_n, b_0 = z - X\beta$.

c) Calculate $p(\sigma^2|\beta^1, \theta^1, \rho^0, z^0, y)$. We carry out a random draw to determine $\sigma^{2(1)}$.

$$\sigma^2|(\beta, \theta, \rho, z, y) \sim IG(\alpha_0, v_0) \tag{4.15}$$

where $\alpha_0 = \frac{n}{2} + \alpha, v_0 = v + \frac{1}{2}\theta'S'S\theta$.

d) Calculate ρ using θ^1 and $\sigma^{2(1)}$ from previous steps. We represent the posterior distribution of ρ as follows:

$$p(\rho|\beta^1, \theta^1, \sigma^{2(1)}, z^0, y) \propto |I_n - \rho W| \cdot exp\left\{-\frac{1}{2\sigma^2}\theta'S'S\theta\right\} \tag{4.16}$$

It is difficult to sample a draw from this distribution. Therefore, we use the Metropolis-Hastings algorithm with a random walk chain to generate draws. Let ρ^{old} denote the previous draw, and then the next draw ρ^{new} is given by

$$\rho^{new} = \rho^{old} + c^*\phi \qquad \phi \sim N(0, 1) \tag{4.17}$$

where c* is called a tuning parameter. The spatial term ρ is restricted $\lambda^{-1}_{min}, \lambda^{-1}_{max}$. Next, we evaluate the acceptance probability as follows:

$$\Psi(\rho^{old}, \rho^{new}) = min\left(1, \frac{p(\rho^{new}|\beta^1, \theta^1, \sigma^{2(1)}, z^0, y)}{p(\rho^{old}|\beta^1, \theta^1, \sigma^{2(1)}, z^0, y)}\right) \tag{4.18}$$

Finally, we set $\rho = \rho^{new}$ with probability $\Psi(\rho^{old}, \rho^{new})$, otherwise $\rho = \rho^{old}$.

e) We sample z^1 draws from a truncated normal distribution using $\beta^1, \theta^1, \sigma^{2(1)}$ and ρ^1 as follows:

$$z_i|(\beta,\theta,\rho,\sigma^2,z_{-i},y) \sim$$

$$\begin{cases} TN_{(0,\infty)}(x_i'\beta+\theta_i,1) & if \quad y_i=1 \\ TN_{(-\infty,0]}(x_i'\beta+\theta_i,1) & if \quad y_i=0 \end{cases} \tag{4.19}$$

where $z_{-i}=(z_1,\cdots,z_{i-1},z_{i+1},\cdots,z_n)$

We now return to step 1 employing the updated parameter values in place of the initial values β^1, θ^1, $\sigma^{2(1)}$, ρ^1 and z^1. On each pass through the sequence, we collect the parameter draws which are used to construct a posterior distribution for the parameters in our model.

4.4 RESULTS AND DISCUSSIONS

4.4.1 The Explanatory Variables

Since our research aim is to investigate individuals' preference regarding water choice at the community level, in this paper we define our sample with two considerations. First, only respondents with more than one option of access to clean water could be taken into account in the sample. And there are 317 respondents. Secondly, only those with complete answers to eight questions requesting demographic data and one question requesting psychological data are included as respondents in the sample. Thus, the sample for this paper is comprised of 257 respondents. Then, the dependent variable is set to 1 for respondents who participate in CBWM – community-based water management (HIPPAM's members and those who get water from public hydrants), and 0 for respondents not participating in community-based water management or for those obtaining clean water from different access (PDAM, individual wells, and others). Table 4.2 shows the number of respondents comprising the sample in the paper.

Table 4.2 Sample of respondents.

Sample of Respondents			
Participating in CBWM	HIPPAM	167	185
	Public Hydrant	18	
Not participating in CBWM	PDAM	18	72
	Individual Well	51	
	Others	3	
Total Sample			257

In detail, 65% of the respondents obtain water from HIPPAM, and 7% acquire water from public hydrants. Meanwhile, PDAM customers account for 7%, respondents who use individual wells account for 20%, and others account for 1%.

Table 4.3 illustrates characteristics of respondents from the aspect of water source, covering 8 demographic data and one psychological datum of the 257 respondents. Groups are comprised of A – respondents who join community-based water management (n = 167), and B – respondents who do not participate in community-based water management (n = 90). In general, data in the sample reflect a situation similar to that with the total 500 respondents.

Table 4.3 Characteristics of respondents from the aspect of water source.

	Gender Male	Age	Family Size	Length of Stay	Water Usage
A	75%	Av.45.87	Av.3.96	Av.30.17	Av.1789.98
B	71%	Av.48.61	Av.4.01	Av.27.42	Av.806.89

	Education		Occupation		Concern for Com. Welfare	
	M.S. & below	J.S. & upper	Agric. & Manu.	Serv. & Unempl.	Agree	Disagree
A	80%	20%	45%	55%	30%	70%
B	72%	28%	24%	76%	39%	61%

	Income						
	0.25	0.75	1.25	1.75	2.25	2.75	3.25
A	26%	40%	19%	8%	2%	3%	2%
B	28%	37%	16%	8%	2%	2%	8%

In addition, with regard to daily water usage, in the questionnaire sheet, we asked respondents to describe their average household daily water usage. According to Peter Gleick (1998), a standard of minimum water requirements per person per day is 50 liters. Since average family size is 4, then we define daily water usage per household as 200 liter/household/day, by multiplying the standard minimum water requirements by the number of family members.

In order to investigate the concern of the people regarding community welfare in the study area, we raised this question: "Do you agree or disagree with the following statement: People here look out mainly for the welfare of their own families and they are not much concerned with community welfare," and asked respondents to choose one of these four categories: Strongly Agree – Agree – Disagree – Strongly Disagree. Then, after we grouped the responses into two classifications - Agree and Disagree – the result showed that the majority of respondents have concern toward community welfare.

Considering the above data, we define explanatory variables as follows:

- *GENDER* : dummy variable which equals 1 if respondent is male.
- *AGE* : age of respondent.
- *FAM* : number of people in each household.
- *LENGTH* : years of living in the area for respondent.
- *WATER USAGE* : dummy variable of daily water usage, categorized as 1 if respondent consumes less or equal to 200 liters/household/day and categorized as 0 if consumption is more than 200 liters/household/day.
- *EDU* : dummy variable which is recorded as 1 if respondent has educational background at the level of middle school or below, and recorded as 0 if high school or above.

- *OCCU* : dummy variable which is coded as 1 if occupation of respondent is agriculture or manufacturing and coded as 0 if service or unemployment.
- *CONCERN FOR COMMUNITY WELFARE* : dummy variable which equals 1 if respondent's answer is *disagree* or *strongly disagree* and recorded as 0 if respondent's answer is *agree* or *strongly agree*.
- *INCOME* : household's monthly income which is divided into 7 items (less than 0.5, 0.5–1.0, 1.0–1.5, 1.5–2.0, 2.0–2.5, 2.5–3.0, more than 3.0 million Rupiah), with medians used for each item.

Table 4.4 shows the standard statistics for each variable.

Table 4.4 Standard statistics.

N= 257	Mean	STDV.	Max	Min
GENDER	0.74	0.440	1	0
AGE	46.56	11.373	75	22
FAM	3.97	1.123	7	2
LENGTH	29.40	18.958	75	0.02
WATER U.	0.65	0.479	1	0
EDU	0.77	0.421	1	0
OCCU	0.38	0.486	1	0
CONCERN C.W.	0.67	0.471	1	0
INCOME	0.98	0.751	3.25	0.25

4.4.2 The Weight Matrix

Finally, in specifying the weight matrix, we reasoned that households with better community ties have the ability to organize a community-based management system. Therefore, we define a spatial weight matrix using the data about community networks. To investigate the participation of respondents in various types of social organizations and informal networks, and the range of contributions given and received, we asked respondents to identify from the 15 following options the type of community group which they join: (1) Religious, (2) Cultural/Social, (3) HIPPAM, (4) PDAM, (5) HIPPA, (6) Ethnic based, (7) Community organization, (8) Finance, (9) Production, (10) Union (labor/trade), (11) Political party, (12) Professional association, (13) Business association, (14) Social movement, and (15) Others (please specify). Table 4.5 shows the number of respondents in each type of community group.

Table 4.5 Number of Respondents in Each Type of Community Group.

	1	2	3	4	5	6	7	8	9	10	11	12	13	14	15
A	155	25	0	83	1	5	21	21	5	1	1	2	1	0	1
B	65	6	4	0	0	0	9	5	0	1	1	0	2	1	1
Tot.	220	31	4	83	1	5	30	26	5	2	2	2	3	1	2

It is evident that the four groups most popular among respondents are (1) Religious, (2) Cultural/Social, (7) Community organization, and (8) Finance, at participation rates of 75%, 21%, 18%, and 9%, respectively, among total responses. Therefore, we focus on these four types of groups and formulate the weight matrix by calculating the social distance between household i and household j as follows:

$$w_{ij}^{g} = \begin{cases} 1 \text{ if household } i \text{ and } j \text{ join the same social group } k \\ 0 \text{ otherwise} \end{cases}$$

$$\mathrm{W}_{ij} = \sum_{k=1}^{4} w_{ij}^{g}. \tag{4.20}$$

The diagonal elements were all set to zero. Next we row standardize the matrix by dividing each element w_{ij} in the matrix by the row sum such that all rows sum to one. The row standardization does not change the relative social interaction among households. Other more complicated weighting schemes are possible, depending on how one wishes to quantify the degree of social interaction among households. For the purpose of this paper, we simply want to account for social interaction effects in the decision to join HIPPAM; therefore any type of social interaction is acceptable.

Table 4.6 Estimation results (Toyomarto).

		Toyomarto(n=159)			
		Probit Model			
Variable	**p.Mean**	**p.Std.Dev.**	**90% credible**	**Interval**	**Geweke**
Constant	3.449	1.389	1.266	5.787	0.318
FAM	–0.734	0.178	–0.363	0.218	0.155
GENDER	–0.058	0.468	–0.838	0.697	0.408
AGE	–0.029	0.021	–0.063	0.005	1.097
EDU	–0.069	0.533	–0.982	0.794	0.402
OCCU	–0.289	0.428	–1.007	0.383	0.236
INCOME	0.252	0.363	–0.326	0.876	1.057
LENGTH	0.024	0.013	0.002	0.046	0.476
COST	–0.004	0.001	–0.005	–0.003	0.390
		Spatial Probit Model			
Variable	**p.Mean**	**p.Std.Dev.**	**90% credible**	**Interval**	**Geweke**
Constant	3.620	1.545	1.212	6.200	1.872
FAM	–0.085	0.204	–0.415	0.246	0.603
GENDER	0.035	0.509	–0.801	0.861	0.607
AGE	–0.031	0.023	–0.069	0.006	1.680
EDU	–0.136	0.579	–1.112	0.788	0.322
OCCU	–0.234	0.485	–1.062	0.561	2.863
INCOME	0.228	0.367	–0.364	0.844	0.499
LENGTH	0.026	0.014	0.003	0.049	1.949
COST	–0.004	0.001	–0.005	–0.003	2.224
	0.128	0.027	0.092	0.177	1.220
	–9.016	5.486	–17.345	0.175	1.203

Table 4.7 Estimation results (Candi Renggo).

		Candi Renggo(n=142)			
		Probit Model			
Variable	**p.Mean**	**p.Std.Dev.**	**90% credible**	**Interval**	**Geweke**
Constant	–1.272	0.749	–2.501	–0.045	1.634
FAM	0.199	0.119	0.006	0.394	1.360
GENDER	0.554	0.351	0.005	1.144	0.003
AGE	–0.026	0.014	–0.048	–0.004	0.787
EDU	0.865	0.340	0.312	1.446	0.740
OCCU	1.059	0.363	0.478	1.661	1.053
INCOME	0.129	0.181	–0.168	0.426	0.059
LENGTH	0.039	0.011	0.020	0.058	0.831
COST	–0.001	0.0002	–0.001	–0.0003	1.147
		Spatial Probit Model			
Variable	**p.Mean**	**p.Std.Dev.**	**90% credible**	**Interval**	**Geweke**
Constant	–1.376	0.800	–2.714	–0.092	0.509
FAM	0.196	0.123	0.002	0.404	0.711
GENDER	0.594	0.370	0.001	1.216	3.406
AGE	–0.026	0.014	–0.050	–0.004	0.519
EDU	0.939	0.362	0.351	1.544	0.578
OCCU	1.157	0.397	0.497	1.819	1.683
INCOME	0.149	0.193	–0.170	0.476	0.544
LENGTH	0.041	0.013	0.021	0.063	0.866
COST	–0.001	0.0002	0.001	–0.0004	0.903
	0.127	0.026	0.091	0.175	0.274
	–5.092	3.222	–10.121	0.067	0.412

4.4.3 Estimation Results

In addition to the spatial probit model estimates, we also estimated a non-spatial probit model which does not include the spatial interaction term θ. Diffuse or conjugate priors were employed for all of the parameters β, σ^2 and ρ in both models. We iterate MCMC algorithm and sample 5000 parameters respectively and set 1000 samples as burn-in. The chain was considered to have practically converged after 1000 iterations based on a diagnostic proposed by Geweke (1992). The last 4000 draws were used to calculate the posterior mean and standard deviation of the parameters. Table 4.6 and 4.7 show the estimation results of each village.

Estimation results are summarized as follows. First, in both the villages, the results indicate very similar inferences would be drawn from the non-spatial probit model versus the spatial probit model. In addition, all of the estimated parameter ρ is negative and insignificant. Therefore, from this result, we cannot confirm the existence of the "social interaction effect" among households.

Second, the estimated parameter LENGTH is positive and significant. This result indicates that the longer the respondents stay in the area of study the higher their preference to join HIPPAM.

Third, we can find that the estimated parameter COST is negative and significant. Thus, even if the price of monthly water usage is quite low, it plays an importants part in determining their willingness or lack of willingness to join HIPPAM from their point of view.

4.5 CONCLUSION

In this paper, we have shown the spatial probit model using the Bayesian estimation method in order to investigate residents' spontaneous collaboration to manage a community-based water supply system. We described the posterior distribution from the Bayes theorem and expressed the MCMC sampling method. Then our approach was applied in an empirical analysis of data from a field survey in Indonesia. From the estimation results, we can say that the length of time living in the area and the price of monthly water usage have an important meaning for respondents when they make decisions about joining a community-based water supply system. However, as for social interaction, we do not yet have a good result for parameter ρ. Therefore, we need to use another approach to get a weight matrix employing geographical neighbors' data through Social Network Analysis in order to complete the previous analysis of demographic neighbors.

In this paper, we focused on the mechanism of residents' spontaneous collaboration to access water and did not take up the social and economic benefits of community-based water supply systems. Needless to say, it is important to clarify the flow of cost and benefit generated by community-based water supply systems. These are issues which remain for future work.

4.6 REFERENCES

Asian Development Bank (2007). *Asian Water Development Outlook 2007: Achieving Water Security for Asia*, Asian Development Bank.

Chib S. and Greenburg E. (1995). Understanding the metropolis-hastings algorithm. *American Statistical Association*, **46**(4), 327–336.

Chiristiaan G., Deepa N., Nyhan J. V. and Woolcock M. (2004). *World Bank Working Paper No.18: Measuring Social Capital An Integrated Questionnaire*. The World Bank, Washington DC.

Gelfand A. E. and Smith A. F. M. (1990). Sampling-based approaches to calculating marginal densities. *Journal of the American Statistical Association*, **85**, 398–409.

Geweke J. (1992). Evaluating the accuracy of sampling-based approaches to the calculation of posterior means. In: J. M. Bernardo, J. O. Berger, A. P. Dawid and A. F. M. Smith (eds.), Bayesian Statistics 4. Oxford University Press, Oxford.

Gilks W. R., Richardson S. and Spiegelhalter D. J. (1996). *Markov Chain Monte Carlo in Practice*, Chapman & Hall/CRC.

Gleick P. (1998). Water in crisis: paths to sustainable water use. *Ecological Applications*, **8**(3), 571–579.

Lesage J. P. (2009). *Introduction to Spatial Econometrics*, CRC Press.

NARBO Annual Report (2007). Networks of Asian River Basin Organizations.

Nationmaster.com (2000). Average years of schooling of adults. http://www.nationmaster.com/country/id-indonesia/edu-education, (accessed 3 April, 2012).

Smith T. E. and Lesage J. P. (2004). A bayesian probit model with spatial dependencies. In: James P. Lesage and R. Kelley Pace (ed.), *Spatial and Spatiotemporal Econometrics (Advances in Econometrics, Volume 18)*, Emerald Group Publishing Limited, pp.127–160.

Statistic Centre Board Indonesia (2005). Badan Pusat Statistik, Jakarta.

World Health Organization and UNICEF (2006). Meeting the MDG drinking water and sanitation target: the urban and rural challenge of the decade, WHO and UNICEF, Switzerland.

Woolcock M. and Narayan D. (2000). Social capital: implications for development theory, research and policy. *The World Bank Research Observer*, **15**(2), 225–249.

Wydick B. (2008). *Games in Economic Development*. Cambridge University Press, New York.

5 Social capital and governance for efficient water management

Hans Westlund

Department of Urban Planning and Environment, KTH Royal Institute of Technology, Drottning Kristinas väg 30, SE-100 44 Stockholm, Sweden. E-mail: hans.westlund@abe.kth.se

Abstract

Social capital is a multi-faceted concept and connected to all human interaction. Therefore, social capital is important in all growth and development policies. No matter which water supply solution is implemented, or the number of users, good governance of the water supply system is necessary, and good governance cannot be achieved without good social capital. A careful combination of top-down and bottom-up methods is necessary, in unique combinations for each case.

Participatory planning is one way to achieve the good governance that produces good social capital, but there has been little experience above the very local, neighborhood or village level. Civil society is often underdeveloped outside the western world. NGOs often represent ideological, economic or donors' interests. This means that they are not a guarantee for "good governance," even if they can play a very positive role in implementing concrete projects. Policies supporting citizens' own engagement, and against corruption, are the best for good governance and social capital that promote efficient water management systems.

Keywords: Social Capital, Planning, Governance, Water Management.

5.1 WATER – AN UNEVENLY SUPPLIED INDISPENSABILITY FOR LIFE

There is probably no substance that is so connected to life as water. When space researchers say that there seems to be water on Mars, it is immediately connected to the possibility that some kind of life exists or has existed there. Without water, no life can exist. One of many reflections of this is that mammals, such as us human beings, consist mainly of water.

Our planet, the earth, is often called "the blue planet," as most of its surface is covered with water. However, salt water is impossible to use for most human purposes, such as drinking and cleaning. Fresh water is the resource that we have the most need for, and about 1.1 billion people lack enough of it. Of these people, 70 percent live in just 11 countries: India, China, Indonesia, Nigeria, Bangladesh, Pakistan, Ethiopia, DR Congo, Vietnam, Brazil and Afghanistan.

It is important to state that the problem of water scarcity is only partly due to a physical lack of water. To a much larger extent, water shortage is a poverty problem, a lack of economic resources to build the structures necessary for water supply. A striking illustration of this is that the average American uses 100 times more water per day than the average African.

Water shortage is a growing problem and there are a number of mutually connected reasons for this. Population growth in developing countries and newly industrializing countries means in itself increased demand for water in a number of areas. Still more important reasons for increased demand are economic development and all the factors connected with it: industrialization, pollution, energy production and consumption, food production, transportation and all other factors associated with higher living standards. Also, it has been argued that climate change alters weather patterns, resulting in drought and floods in areas that formerly did not suffer from such problems. Together, these factors lead to increased competition for scarce water, with ever increasing uneven access to water as a result.

5.2 THE MULTIPLE IDENTITIES OF WATER

From an economic-political view, the question arises as to whether water is a public good, a commodity or a human right. The answer is that water has characteristics of all three: a public good, a commodity and a human right. The problem is that these multiple identities of water often come in conflict with each other.

Regarding the issue of water as a public good, it must be stated that water cannot be defined as a perfect public good. Public goods are "an exception to the free-market system marked by two characteristics. First, if one person consumes public goods, the amount available remains the same. Second, once public goods are available, no one can be stopped from consuming them for free. An example is the protection provided by police, fire departments, and the military" (www.businessdictionary.com).

Instead, like transportation and communication networks, education and other goods provided mainly by government, water is a semi-public good. Government is responsible for an efficient supply, but does not have to be the supplier. In contrast to pure public goods, access to semi-public goods is regulated by law, etc., or by fees. If a semi-public good is taken for a public good, it might result in the "tragedy of the commons"; that is, unrestricted individual use degrades, damages or destroys the resource. This is what has happened with water in many places.

From a pure and strict economic point of view, just as with land and what is on and under the earth's surface, water is a commodity, among others, that is subject to market transactions. However, in the pure economic point of view, there are no considerations of human and social issues. This is the reason that societies develop institutions other than the market. The pure market needs regulations and laws. People need semi-public goods and other services, of which the pure market often fails to produce a sufficient supply. Being a semi-public good, water is a commodity with a special position.

Both these examples show the importance of correct definitions, both in theory and in practice. Wrong definitions of a resource might lead to the incorrect use of it, with devastating results.

The third identity of water is that of a human right. Human rights are often defined in accordance with United Nations' declarations on civil and political rights and economic, social and cultural rights. In November 2002, the United Nations Committee on Economic, Social and Cultural Rights issued a non-binding comment affirming that access to water was a human right:

"The human right to water is indispensable for leading a life in human dignity. It is a prerequisite for the realization of other human rights" (United Nations, 2002). Still, what should be included in human rights is a political question. There is no binding international agreement that water should be included among human rights. Of course, this does not say that government should not be responsible for the provision of water for all citizens.

5.3 THREE ALTERNATIVES FOR WATER MANAGEMENT

In the discussion on how water management should be sustainably organized, in principle three main opinions have crystallized:

- The "tragedy of the commons" must be avoided through *privatization* (IMF, World Bank)
- *Citizens* should be able to plan and manage water supply systems by themselves (Ostrom, 1990)
- *Government* is responsible for the provision of a number of semi-public goods since the market fails to provide the amount or standard that the citizens want (Traditional European view, supported by e.g. Suleiman, 2010)

The first of these views often uses arguments originally presented by an ecologist warning about population growth (Hardin, 1968). According to Hardin, individuals, acting rationally, can destroy a common resource by over-exploitation. His most famous example is that of one herder who receives all the benefits from an additional grazing cow, while the damage to the common pasture is shared by the entire group. This example has often been used as an argument for the privatization of common resources, e.g. water. However, it should be noted that this example can also be used as an argument for governmental regulations and control or management by government.

The most prominent advocate of the second view is Elinor Ostrom, Nobel laureate in Economics 2009. The Royal Swedish Academy of Sciences (2009) cited Ostrom for her analysis of economic governance, saying her work had demonstrated how common property could be successfully managed by the groups using it. Her "research brought this topic from the fringe to the forefront of scientific attention (...) by showing how common resources can be managed successfully by the people who use them, rather than by governments or private companies." Ostrom's work has challenged conventional wisdom, showing that *common resources can be successfully managed without government regulation or privatization*.

The third view is that water supply systems are the government's responsibility. As 90 percent of water systems are publicly owned and managed, this seems to be the main opinion in the world.

Government manages water supply by monopoly or by contracts with private companies. However, all governments are not equipped to manage water, from up-front investments to maintenance of infrastructure to distribution. Multiple levels of competent governance are required for effective water management. This is sometimes a problem in developing countries.

5.4 PROBLEMS OF GOVERNMENT IN MANY DEVELOPING COUNTRIES

Many developing countries are facing a number of interconnected problems that have an impact on their ability to secure a functioning water supply for their citizens. One such problem is the lack of capital for investment in and maintenance of water infrastructure.

Water pipelines and sewage systems are examples of capital-heavy infrastructure, and in a developing country the competition for investment between various projects is particularly strong. The need for economic development often means that resource-exploiting investments take priority over people's need for water. Other problems mentioned in this discussion are lack of competence, inefficient administration and corruption.

Governments' weaknesses and inefficiency comprise the main argument in favor of water privatization. Privatization of water ownership and management has been strongly supported by the World Bank and the International Monetary Fund. Some form of privatization has been implemented in areas of a rather large number of countries: Argentina, Armenia, Bolivia, Brazil, Chile, Colombia, Cote d'Ivoire, Cuba, Ecuador, England/Wales, Gabon, Ghana, Honduras, Indonesia, Jordan, Morocco, Philippines, Senegal, South Africa, Tanzania and Uruguay. Privatization has taken place mainly in the capital or other metropolitan areas.

Where water privatization has been implemented, the main experience seems to have been improved physical access but higher prices. There are also examples of violent protests, as in the case of Bolivia. However, of the water systems being privatized during the 1990s, 84 percent remained under private control in 2007. This seems to be an indication that privatization has been accepted in most areas in which it has been implemented. A possible interpretation of this is that in most areas, in spite of the higher costs, the private water supply has functioned better than the public alternative did.

What, then, are the water supply alternatives for government?

The most common alternative in developed countries is an own monopoly on water supply. Such monopolies are more common in urban areas than in semi-urban and rural areas, but in e.g. Sweden this local government monopoly is functioning also in small villages. The very opposite alternative for government would be to take no action at all and let private, individual alternatives develop without any regulations. However, this laissez-faire policy does not exist anywhere – something that must be interpreted as signifying that water cannot be treated in the same way as other commodities. This means that the only realistic alternative to government monopoly is some form of privatization of the former government monopolies.

Privatization is still not just "privatization." In fact, as can be seen in Table 5.1, various types of privatization can be listed on a scale from total to time-limited privatization.

No matter which alternative government chooses, monopoly or different forms of privatization, good governance is the key to a water supply system that works well at low cost.

Table 5.1 Types of privatization of water management, in order from total to time-limited privatization.

Types of privatization
Total privatization (parts of England & Wales)
Management contract with private company that runs the system
Lease contract under which assets are leased to the private operator, investment financed by government
Concession, private operator responsible for running the whole system, including investment. Assets remain publicly owned
BOT (Build-Operate-Transfer), for carrying out specific new investments. Private partner constructs the system and runs it for a number of years, during which payment is received, before it's handed over to government

5.5 WHAT IS "GOOD GOVERNANCE" IN WATER MANAGEMENT?

To begin with, governance as such can be summarized in the following points:

- Relationships where government is involved, expressed in various partnerships and networks with the private sector and the civil society
- Societal processes for achieving collective action in public affairs
- A tool to allocate, implement and regulate social and political actions

Water governance can be described as "the range of political, social, economic and administrative systems that are in place to develop and manage water resources and the delivery of water services, at different levels of society" (Global Water Partnership, 2003). The European Commission in 2001 set up six principles for "good governance" that can be applied to water governance:

- Openness and transparency
- Inclusiveness and communicativeness
- Coherent and integrated policy
- Equitability
- Accountability
- Efficiency

An important aspect of this discussion is that government's initiatives to a very large extent take place only in urban areas. Urban areas are denser than rural ones, and this means lower investment costs per user in cities than in the countryside. Thus, governmentally organized water supply is less frequent in the countryside and is less likely the sparser the settlement pattern is. Most often, the countryside's alternatives are to find private, individual solutions, to raise demands on government to extend its water supply to rural areas, or to organize its own water supply system. Of these alternatives, we will discuss the last one more in detail.

5.6 SELF-ORGANIZED SYSTEMS FOR MANAGEMENT OF COMMON RESOURCES

There are many examples all over the world of organizations created to manage a common resource: jointly owned land or water, irrigation systems, apartment blocks, etc. All these examples are living illustrations that the tragedy of the commons need not happen. How do these examples avoid the problems of free-riding and/or over-exploitation?

Many of these self-organized management systems seem to have emerged independent of each other. Elinor Ostrom has summarized the experiences from these systems into a number of design principles for sustainable local management systems:

- Clear definition of membership/user of system
- Rules for what the member receives and is expected to contribute
- Possibility for members to change management rules
- Possibility for members to choose their own supervisors/managers, who are responsible to the members

- Possibility to issue sanctions against those who break the rules
- Access to local and cost-efficient arenas for conflict resolving
- Legal support for self-organizing management

5.7 THE IMPORTANCE OF SOCIAL CAPITAL

Design principles for local management systems are good, but they must be maintained and sustained. "Good governance" is not achieved by the stroke of a magic wand. Sustainable, efficient water management systems – self-organized, public or private – need networks for information exchange, negotiations and trustful management; i.e., they need *social capital.*

Social capital can be defined as non-formalized *networks* that are used by the networks' nodes/actors to distribute *norms, values, preferences* and other social attributes and characteristics. In this way, social capital forms an informal counterpart to formal institutions (rules) and organizations. Another way to describe this is that social capital is a type of infrastructure (with nodes and links) and "traffic" (of information, values, etc). This means that not only does the quality of the social links and values matter, but the quantity matters as well (Westlund, 2006).

The self-organized systems for water management are most often rural phenomena. What aspects of social capital are important in such rural development processes? Table 5.2 summarizes a number of aspects of social capital that are of importance.

5.7.1 Historical Social Capital

The existing stock of social capital is crucial at the initial stage of an activity's local development process. The start-up of development processes is affected by existing social networks, norms and values. These factors influence the local community's acceptance of new ideas and beliefs in the feasibility of achieving change.

5.7.2 Confidence as Social Capital

In the start-up phase, individuals' own social capital is crucial to their ability to mobilize resources for a novel initiative. People's confidence in the promoters and leaders is vital.

5.7.3 Networking as Social Capital

Social capital is generated through interaction between individuals and groups. The creation of networks and cooperation platforms supports the building up of new social capital. However, excessively narrow interaction among a group of individuals can create social capital in which considerations of protection and control lead to the exclusion of other people - a form of social capital which also can have a negative influence on local development initiatives.

5.7.4 Learning as Social Capital

Where the existence of "good" social capital is obvious, there should (according to theory) be openness to learning processes and new knowledge. Learning functions as a catalyst for entrepreneurship and development. However, a prerequisite for learning is an understanding by individuals of their need to reinforce their expertise.

Table 5.2 Phases of local development and their interaction with different aspects of social capital.

Phases of local development	Aspects of social capital	Comments
Background of and reasons for the initiative	Historical social capital	Directly needed in order to establish e.g. Leader-partnerships.
Start-up and leadership	Confidence as social capital	Within the entrepreneurs who believe in the business idea or the leading figure at the local community centre.
Aims and flexibility	Learning as social capital	Having competence to adjust aims to the changing situation in a flexible way. Strategically seeking for knowledge outside the region and from other parts of Europe.
Sustainability, business soundness and entrepreneurial merits	Economy-related social capital Networking as social capital	Local platforms and structures support enterprises and local initiatives economically with public and private finances. Network between entrepreneurs and local individuals in order to improve local growth and competitiveness.
Mobilization and commitment of actors	Mobilizing role of social capital	The association expands its responsibilities due to the mobilizing of individuals.
Results obtained	Networking as social capital Economy-related social capital	Platform structures and financial capital organized in a bottom-up approach where decisions made on local level build and maintain social capital.

Source: Fröbel and Westlund (2009)

5.7.5 Economy-Related Social Capital

Joint action, for example in local partnerships, joint ownership, cooperatives or other local economic relationships, is based on a social capital aiming at utility for all participants.

Thus, joint action relies on the presumption that "what is good for the village or neighborhood is good for everybody."

5.7.6 The Mobilizing Role of Social Capital

Social capital can act as a mobilizer when less active local development processes suddenly come to life and in a short time engage the local community.

5.8 SOCIAL CAPITAL IN THE GOVERNANCE OF *URBAN* WATER SUPPLY

Organizing a functioning water supply in cities is in general a much more complex task than it is in the countryside; more actors are involved and thereby more conflicting interests. For this reason, self-organizing water management systems, with their special form of social capital, are very rare in cities. Instead, there is a need to highlight other aspects of social capital. Three general types of social capital within and between groups of actors/stakeholders are of importance: *bonding* social capital, *bridging* social capital, and *linking* social capital.

5.8.1 Bonding Social Capital

Bonding social capital consists of horizontal and vertical links contributing to the internal solidarity of a group of actors, e.g. water users, NGOs, governmental bodies, well-established Public-Private Partnerships, etc. Bonding social capital is an absolute necessity for the cohesion of a group of actors; without it, the group would cease to exist. But too strong bonds might mean unwillingness to cooperate outside the own group, or to listen to others' arguments, etc. Therefore, bonding social capital needs to be supplemented by other aspects of social capital.

5.8.2 Bridging Social Capital

In the literature, bridging social capital is defined as horizontal, external links between actors of different groups. As governance is about involving various actors in decisions and implementations, bridging social capital is a key ingredient for achieving "good governance."

Moreover, bridging social capital is necessary for local/regional/societal cohesion, for information exchange, knowledge spillovers, establishment of new Public-Private Partnerships, etc. In addition, it is a requirement for participatory planning and participatory development.

5.8.3 Linking Social Capital

Linking social capital can be described as vertical links between actors at different levels of society, e.g. citizens, government, employers, employees, local organizations, and national decision makers. Also, linking social capital is necessary for participatory planning and participatory development.

5.8.4 Social Capital and Governance

The relationship between social capital and governance is mutually reinforcing. Good governance improves bridging and linking social capital. Good bridging and linking social capital improves good governance, i.e. openness and transparency, inclusiveness and communicativeness, coherent and integrated policy, equitability, accountability and efficiency.

5.9 SOME CONCLUSIONS

Social capital is a multi-faceted concept and connected to all human interaction. Therefore, social capital is important in all growth and development policies. No matter which water supply solution is implemented, or the number of users, good governance of the water supply system is necessary, and good governance cannot be achieved without good social capital. A careful combination of top-down and bottom-up methods is necessary, in unique combinations for each case.

Participatory planning is one way to achieve the good governance that produces good social capital, but there has been little experience above the very local, neighborhood or village level. Civil society is often underdeveloped outside the western world. NGOs often represent ideological, economic or donors' interests. This means that they are not a guarantee for "good governance," even if they can play a very positive role in implementing concrete projects. Policies supporting citizens' own engagement, and against corruption, are the best for good governance and social capital that promote efficient water management systems.

5.10 REFERENCES

Business Dictionary. www.businessdictionary.com (accessed 2013-06-26).

Fröbel L. and Westlund H. (2009). Social capital as a driving force in local development and social economy entrepreneurship - a qualitative study based on Swedish examples. In: Social Capital and Development Trends in Rural Areas, **Vol. 4**, K. Kobayashi, H. Westlund, T. Tamura and H. Jeong (eds.), MARG, Kyoto University, Kyoto, pp. 50–66.

Global Water Partnership (2003). http://www.gwp.org/Global/About%20GWP/Annual%20Reports/GWP_in_action%202003.pdf (accessed 2013-06-26).

Hardin G. (1968). The Tragedy of the Commons. *Science*, **162**(3859)(December 13, 1968), pp. 1243–1248.

Ostrom E. (1990). *Governing the Commons: The Evolution of Institutions for Collective Action*. Cambridge UP, Cambridge.

Sulieman L. (2010). *Water Governance in Transition*, PhD thesis, Department of Urban Planning and Environment, KTH Royal Institute of Technology, Stockholm, Sweden.

The Royal Swedish Academy of Science (2009). http://www.nobelprize.org/nobel_prizes/economic-sciences/laureates/2009/press.html (accessed 2013-06-26).

United Nations (2002). United Nations Committee on Economic, Social and Cultural Rights. General Comment 15. http://www.unhchr.ch/tbs/doc.nsf/0/a5458d1d1bbd713fc1256cc400389e94/$FILE/G0340229.pdf (accessed 2013-06-26).

Westlund H. (2006). *Social Capital in the Knowledge Economy: Theory and Empirics*. Springer, Berlin, Heidelberg, New York.

6 Trust formation in collaborative water management

Kiyoshi Kobayashi[1], Tsuyoshi Hatori[2], and Hayeong Jeong[3]

[1] Graduate School of Management, Kyoto University, Katsura, Nishikyo-ku, Kyoto, 615-8540, Japan. E-mail: kkoba@psa.mbox.media.kyoto-u.ac.jp

[2] Department of Civil and Environmental Engineering, Ehime University, 3 Bunkyo-cho Matsuyama, Ehime, 790-8577, Japan. E-mail: hatori@cee.ehime-u.ac.jp

[3] Graduate School of Management, Kyoto University, Katsura, Nishikyo-ku, Kyoto, 615-8540, Japan. E-mail: hayeong@psa2.mbox.media.kyoto-u.ac.jp

Abstract

A collaborative approach is currently considered key to improving water resources management. This chapter focuses on two conditions necessary for collaborative regimes of water management: common language and trust. Through a subjective communications game, we develop a theoretical model on the truster-trustee relationship between a policy maker and an individual (median voter) involved in a water-related project. The model shows that differences in the cognitive schemes of the two parties prevent them from communicating with each other, and then reduce the likelihood of trust formation. The analysis of the model indicates the importance of attaining shared descriptions by formulating a logic model regarding the correspondence between two different languages. Our analysis also shows that the policy maker might face a trade-off between learning an optimal solution and gaining the trust of an individual, resulting in the failure to realize the optimal project for water management. We discuss the roles of third party reviews in achieving language learning and trust formation simultaneously.

Keywords: collaborative water management (CWM), language learning, subjective game, trust

6.1 INTRODUCTION

Management of water resources, including reliable provision of safe drinking water, proper sanitation, development of irrigation systems, conservation of ecosystems, and flood control, is a major regional issue commonly embedded in social, economic and political interactions across a wide variety of stakeholders (Ferreyra *et al.* 2008; Sneddon, 2002). Water problems are usually of direct interest to the entire population of a region and are discussed from a diverse range of stakeholders' views and perspectives. Because of the scarcity of water resources, the demands of all stakeholders for water supply cannot always be adequately satisfied. Water is a limited resource for which the various demands of stakeholders are more or less mutually incompatible or even conflicting. Accordingly, water problems must be dealt with in an environment of high uncertainty and ambiguity, with the complex interconnections between socio-economic and natural environmental factors causing water managers to struggle to find clear solutions.

In response to these challenges, there has been a demand for more participatory, collaborative, and empowering approaches, where the search for solutions to water problems and the adoption of these solutions are carried out interactively by regional agents including but not limited to government or water managers. Integrated water resources management (IWRM) is one of the major concepts that have emerged during the last two decades in the policy domain to promote the coordinated and collaborative development and management of water resources (Biswas, 2004; Hering and Ingold, 2012). Such a collaborative approach (collaborative water management, hereafter CWM) encourages various actors and stakeholders in public and private sectors and civil society to participate in water-related decision processes, engage in much discussion, and come to an agreement.

For CWM to succeed, a variety of interactions across a wider range of stakeholders must be appropriately governed through good order and workable arrangements. Indeed a collaborative approach creates patterns never seen before of social interactions between multiple sectors. Due to difficulties with regard to building collaborative bridges between different sectors, CWM might fail in providing sufficient water services to stakeholders (Margerum, 1999). This study focuses on two conditions on which the governance of integrated regimes for CWM lies: *common language* and *trust*.

First, language structures provide regional actors with interpretive schemes by which they make sense of what others say and do (Kobayashi and Hatori, 2006). They are a set of "category" or "system of relevance" applied in the sustaining of regional communication networks. These interpretive schemes, however, are not fixed unambiguously. The meanings each actor assigns to others' remarks are open to different interpretations. Especially, decisions on water-related policy often demand sophisticated and complicated judgments. It is difficult for people who lack technical knowledge to understand what water experts and professionals say and do. The cognitive dissonance between professionals and non-professionals might become a severe obstacle to mutual communication. To overcome the problem of miscommunication, a common language must be shared among regional actors.

Second, the presence of trust influences the operation of regional water systems (Hatori and Kobayashi, 2011). In particular, policy makers are often confronted with difficulties in convincing regional members of the benefits of policies. Even if water planners propose an innovative watershed project, such a project might not be realized without community acceptance. Faced with political constraints, planners must be primarily concerned with gaining the trust of stakeholders. A relationship of trust, however, cannot always be sustained spontaneously, especially when there is the possibility of asymmetric information between policy makers and other people, causing the problem of moral hazard, e.g. opportunistic behavior and manipulation by policy makers. To obviate this problem, mechanism designs are needed that provide individuals with a rationale to distinguish policy makers as trustworthy.

We address the two conditions by developing a model of the interrelated policy-making processes in CWM of a policy maker and an individual (median voter). We first present a model that highlights a difference in languages between the two players. For this, we propose a subjective communication game, in which players with different languages communicate with each other. In doing so, we show how the difference in languages leads to a communication failure. Then ways for overcoming the miscommunication are discussed. Second, we formalize a model to describe trust formation between the policy maker and the individual. Third, in order to examine interrelated language learning and trust formation, we develop a model of repeated interplays between the two players. Our analysis shows that the policy maker might face a trade-off between learning an optimal solution and gaining the trust of the individual; learning by trial and error might interfere with trust formation, and vice versa. Finally, we analyze the role of third party reviews as institutional arrangements to deal with the trade-off. We point out that the existence of an interpreter who verifies the correspondence between different languages is key to trust formation.

The remainder of this chapter is structured as follows. Section 6.2 presents a model to describe a communication process between a policy maker and an individual and explores the effects of a difference in languages on the result of the communication. As explained above, the model is developed as a subjective game. Section 6.3 first introduces the theory of the trust game by Bacharach and Gambetta (2001). Then we present a subjective trust game, in a framework in which the policy maker and the individual have different cognitive schemes for the same decision problem. Our approach enables us to analyze the effect that the difference in conceptual schemes has on trust formation between them. Section 6.4 presents a model to explore how language learning and trust formation can be achieved simultaneously. Section 6.5 develops a policy game of third party reviews based on the model presented in section 6.4, and examines the performance of review systems.

6.2 PRIVATE LANGUAGE AND COMMUNICATION

6.2.1 A Theory of Communication

Communication, one of the most complex and strategic activities of human beings, may have limited effectiveness. Farrell (1993) identifies two problems which limit the effectiveness of communication. The first problem, which has been widely analyzed by game theorists, is that of *credibility*: 'communication cannot work well when there are incentives to lie' (Farrell, 1993, p.514). Crawford and Sobel (1982) first show formally how information can be transmitted between rational agents through costless messages or *cheap talk*, given the settings of mixed motives in which agents' interests are partly common and partly conflicting. Crawford and Sobel demonstrate that the informativeness of equilibrium plays is limited by the degree to which agents' interests coincide. But still in the standard cheap talk model, an informed agent (sender) cannot deceive an uninformed agent (receiver) in equilibrium, and thus cheap talk can be credible even if not perfectly. Lupia and McCubbins (1998) argue that this result follows from the assumption that both players "know one another well enough to see lies coming in advance" (Lupia and McCubbins, 1998, p.49). Lupia and McCubbins (1998) remove this assumption by modeling an incomplete information game in which a receiver is uncertain as to the type of sender. They demonstrate that deception can occur in equilibrium of the model, implying that communication cannot be credible.

The second problem, which is also widely recognized but has not yet been embodied in formal modeling, is that of meaning: "meaning cannot be learned from introspection" (Farrell, 1993, p.515). Farrell (1993) argues that a message may gain its meaning in one of three ways. First, the meaning of a message can be fixed through the way in which it is used. This idea comes from Wittgenstein (1957): "meaning is use." In game-theoretical settings, the meaning of a message used in equilibrium is fixed by Bayes' rule. Second, the meaning of a message can be determined by introspection. Standard refinement literature in game theory shows that people's introspection puts restrictions on how out-of-equilibrium messages are interpreted (Cho and Kreps, 1987). But such introspection may not work in cheap-talk games, because payoff-irrelevant messages in out-of-equilibrium may hardly restrict players' beliefs. Finally—and this is stressed by Farrell as the key element which plays a stronger role in cheap-talk games—a message phrased in a preexisting rich language may have a focal meaning. Farrell argues that when there is a rich, common language, both the expected message (equilibrium message) and the unexpected one (out-of-equilibrium message or neologism) may be comprehensible, because their literal meaning is common knowledge. As Rabin (1990) says, "a common language consists of (1) a meaningful vocabulary and (2) a common understanding among agents that it is appropriate to interpret statements according to their literal meaning" (Rabin, 1990, pp.145–146). If agents share such a common language, meaningful communication can occur.

6.2.2 The Problem of Private Language

As seen above, Farrell (1993) argues that meaningful communication can be predicted under the assumption of a common language. Farrell's common language assumption, however, cannot always be guaranteed. Especially, stakeholders' perceptions of water resources and its management are often highly heterogeneous (van Bueren *et al.* 2003). Accordingly, water management issues may not be phrased in terms of a common language, but rather in diverse language structures that provide individual stakeholders with their own understanding. For local water managers, for example, each management alternative may be interpreted in terms of a "means" to tackle water problems (Roberts and Scapens, 1985). Local managers generally have an interest in a specific action for water management and its technical conditions, such as scientific criteria for water quality. In contrast, urban residents not directly involved in the management practices may view each alternative in terms of its "outcome," which represents what influence on his or her realms of imagination the alternative has. Urban residents' interest might be in the taste of drinking water or in the protection of ecosystems. Indeed, it is likely that the difference in interpretive schemes by which stakeholders understand and describe water problems will lead to severe conflict among them. In this section, we focus on the problem of such a disparity in language structures among stakeholders.

In the field of sociology, several empirical studies have showed that differences in the way people describe their problem situations prevent them from solving the problem cooperatively (c.f. Spector and Kitsuse, 1977). In the field of psychology, several theoretical and experimental results regarding how subjects describe the world have been accumulated. Festinger's cognitive dissonance theory (1957) suggests that we have an inner drive to hold all our attitudes and beliefs in harmony and avoid disharmony. A variety of experiments have demonstrated that people cannot correctly perceive a situation because of their cognitive bias and stereotype. A theory of bounded rationality implies that people usually have reasons for what they do, and they evaluate the rationality of their behavior on the basis of the reasons (Selten, 1990; Simon, 1995; Rubinstein, 1998). The experimental results of Kahneman and Tversky (1984; 1979) have revealed the flaws in theories of decision making that do not account for framing effects. The framing effect has been the object of numerous studies conducted in a variety of contexts (c.f. Piñon and Gambara, 2005). The framing effect can be attributed to people's limited cognitive capacity and scarcity of information about the complexity of decision problems.

In contrast, standard game-theoretic analysis starts from the mathematical description of games, and the meaning of the description is assumed to be common knowledge among players as a focal meaning. This is because of the strong presumption that standard game theory has imposed; that is, rational players should not be concerned with how each player describes the game. Some game theorists, however, have challenged this assumption. For example, Kreps (1990) has argued that "in choosing his actions in the short run, the individual builds a model of his choice problem, a model which is typically a simplification or a misspecification (or both) of the 'true situation'" (Kreps, 1990, p.152). More recently, Camerer (2003) has suggested that "players presumably construct internal representations that might barely resemble (games in the form of) matrices or trees" (Camerer, 2003, p.229). Devetag and Warglien (2008) present experiment results showing the limits of a subject's cognitive ability to represent the game correctly.

Several recent studies have proposed game models in which players have different mental schemes to describe the game, i.e., different languages, labels and framings (Bacharach and Stahl, 2000; Casajus, 2000; Sugden, 1995). These studies have demonstrated that a difference in game descriptions between players has a significant influence on equilibrium plays. They mainly focus on how players describe their strategies and assume that the descriptions do not have a substantial influence on players' payoffs. Kobayashi and Hatori (2006) model cheap-talk communication for forest management decisions as a subjective communication game, in which two parties who describe the game differently communicate with each other and decide on one management alternative as a collective choice. In contrast to previous researchers, they propose the concept of *subjective payoffs*, which depend on players' language. In what follows, we will briefly introduce the model of the subjective communication game proposed in this previous research and discuss its implications.

6.2.3 The Subjective Communication Game

Let us consider a regional decision-making process for CWM, in which a policy maker and an individual (median voter in the region) must decide whether or not to undertake a project and if so, which alternative from $n+1$ alternatives to select. It is assumed that the project is carried out only if the two parties come to an agreement. Through a communication process, the policy maker recommends one alternative, and the individual decides whether to accept or reject the alternative. If the individual accepts the alternative, the alternative is selected for the region's project. If she rejects the alternative, the status quo is maintained. In what follows, we impose a strong assumption that the policy maker and the individual have common interests. Even in this context, however, the cognitive dissonance between them might prevent them from finding the optimal alternative. The model shows that communication between the two players with different cognitive schemes ends up in failure.

The property of each alternative is represented in two different languages. On the one hand, the policy maker describes each alternative as a specific action associated with carrying out a project. He might evaluate each alternative on the basis of scientific knowledge about technical conditions regarding the alternative. On the other hand, the individual views each alternative from the standpoint of outcomes that represent what influence the project has. She might be concerned about the matter of environmental preservation from a global standpoint. Let the language that the policy maker uses be called "*action labels*." A set of action labels is represented by $\boldsymbol{A} = (a_0, a_1,..., a_n)$, where a_0 denotes the status quo. Similarly, let the language that the individual uses be called "*outcome labels,*" which are represented by $\boldsymbol{O} = (o_0, o_1,..., o_n)$. Also, o_0 means the status quo. Both players recognize each alternative by means of their own labels, and they do not know how the other player describes the alternative. The model assumes that there is a one-to-one correspondence between the two alternative-labels, and the true correspondence is denoted by one-to-one mapping (function) φ_t; $\varphi_t(a_i) = o_i$ $(i = 0,..., n)$.

Thus, each player recognizes alternatives on the basis of his or her languages (alternative-labels). In other words, the two players play different subjective games, although they communicate with each other around the same decision problem. The model assumes that each player has his or her subjective payoffs in each subjective game. The individual's most preferred outcome is denoted by an ideal outcome o_d, where $d \in [1,..., n]$. The individual privately knows which outcome-label of $[o_1,..., o_n]$ coincides with her ideal outcome o_d. The action that corresponds to the ideal outcome o_d is represented by $a_d = \varphi_t^{-1}(o_d)$. Since the policy maker and the individual are assumed to have common interests, a_d is regarded also as the policy maker's most preferred action, i.e., his ideal action.

Figure 6.1 shows the structure of the subjective games of each player. The left side of this figure denotes the subjective game of the policy maker, while the right side denotes the subjective game of the individual. Here, each game is described in the different languages that each player possesses. This game begins with a move by nature. Nature's move n determines which alternative corresponds to the ideal alternative a_d for the policy maker and o_d for the individual, respectively. Any assignment of the ideal alternative across n alternatives is equally likely. After nature's move, the policy maker recommends one alternative from n alternatives. In the figure, a_{-i} (o_{-i}) means an alternative other than a_i (o_i). Then the individual decides whether to accept or reject the alternative offered by the policy maker. The payoffs of both players depend on the selected alternative. In the case where the ideal alternative is chosen, both players obtain the payoff $x > 0$. In the case where an alternative other than the ideal one is chosen, both players get $\underline{x} < 0$. If the status quo is realized, they get payoff 0.

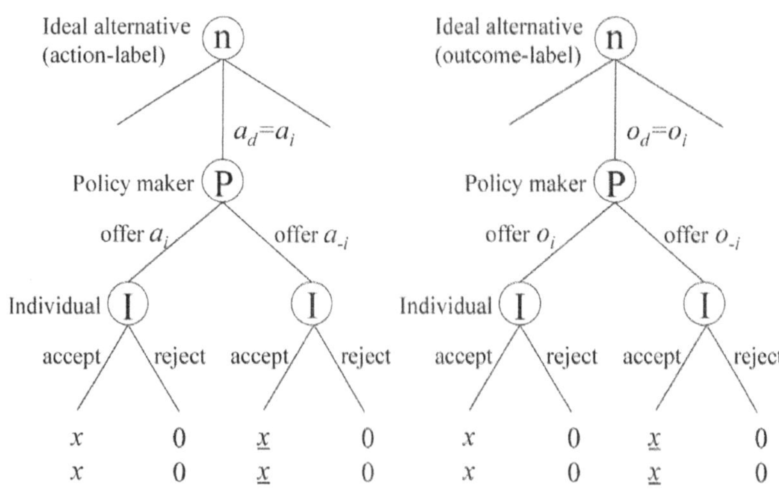

Subjective game of policy maker Subjective game of individual

Figure 6.1 Subjective communication games.

The policy maker who is uncertain about the ideal action a_d possesses a prior belief α_i that represents a subjective probability whereby he expects that action a_i is coincident with a_d. It is assumed that before the game starts, nothing is known to the policy maker about the ideal action a_d, and that he has *ignorance prior*, $\alpha_i = 1/n$, assigned to the event in which each alternative is a_d (Fox and Rottenstreich, 2003). The policy maker offers an action to the individual and can draw a cue for the ideal action a_d from the individual's responses to his offers. After receiving the responses of the individual, the policy maker can update his beliefs about the ideal action a_d.

On the other hand, after receiving the offer from the policy maker represented by the action label $a_i (i = 1,..., n)$, the individual who is uncertain about the true function φ_t anticipates the outcome that corresponds to a_i based on a subjective belief $g(\varphi)$. Let Φ denote a set of possible correspondences between the two labels. The belief $g(\varphi)$ represents a subjective probability distribution over the set Φ that the individual assigns to the event in which a mapping φ is the true function φ_t.

As a solution concept, we analyze the strategy profile reached as a result of iterative pre-game communication where the policy maker recommends an action and the individual replies to the recommendation repeatedly. In this model, there are two equilibria: 1) an alternative is selected through a communication process, and 2) there is no substantial communication, and the status quo is maintained. In case 1, the action-label alternative a_{i*} is selected. The alternative a_{i*} satisfies the following condition:

$$EU(a_{i^*}) = \sum_{\phi \in \Phi} U^E(\phi(a_{i^*}))g(\phi) > 0 \tag{6.1}$$

where $U^E(\varphi(a_i))$ denotes the individual's subjective payoff in the case where outcome $\varphi(a_i)$ is realized in her subjective game.

In case 1, the policy maker supposes that action a_{i*} is coincident with the ideal action. Similarly, the individual perceives that the alternative is better for her and accepts it. The alternative, however, does not necessarily correspond to a_d (or o_d); accordingly, it is possible that the alternative less advantageous to both the individual and the policy maker will be selected. On the other hand, case 2 is realized if there are no alternatives that satisfy condition (6.1). In this case, the individual rejects any alternative the policy maker recommends. Lacking any common points of reference with the policy maker, the individual in case 2 cannot at all decipher the correspondence between the two alternative-labels. As a result, neither player engages in any substantial communication, and eventually the status quo is maintained.

6.2.4 The Importance of Common Language

The model shows that communication between two players with different private languages brings inefficient results: 1) an alternative disadvantageous to both stakeholders is selected, or 2) the status quo is maintained without substantial communication. In the model, the individual does not discern the outcome corresponding to an action the policy maker recommends. Consequently, the individual might accept an action unfavorable to herself, or reject any alternatives. In order for both players to choose, through communication, an alternative beneficial to both, it is required that the correspondence between two different labels be specified, and that both players share their different languages. Let us consider an ideal situation, in which the *logic model* $\varphi_t(a_i) = o_i$ $(i = 0,\ldots, n)$ is common knowledge, and the descriptions of alternatives are shared between two players (McLaughlin and Jordan, 1999). In this context, the individual can distinguish, from among several actions, the one that is in her own best interests. Consequently, by being able to identify through communication the alternative that is better for both players, the players can come to an agreement on the alternative.

In order for a policy maker to carry out a project, it is necessary to specify the execution procedure involving technical considerations, which is represented by action-labels. However, unless the individual who describes the decision problem with more general and virtual language can understand the technical terms, she will not accept the alternative. For the achievement of an agreement on the alternative that is better for both players, common knowledge of the private languages must be realized. The result in this section suggests that the logic model plays an important role in building a common understanding of the project, fostering the sharing of ideas, and facilitating communication.

6.3 TRUST FORMATION IN COLLABORATIVE WATER MANAGEMENT

6.3.1 The Concept of Trust

There has been extensive research in sociology, psychology and economics analyzing various aspects of trust in a vast range of contexts. In a broad sense, the concept of trust is defined as "expectations about others and society". Luhman (1979) characterizes trust as a mechanism that reduces complexity and enables people to cope with the high levels of uncertainty and complexity of contemporary life. According to this definition, trust involves expectations about the persistence and fulfillment of both natural and moral social orders (Barber, 1983). More specifically, Dasgupta (1988) defines trust as "expectations about the actions of other people that have a bearing on one's own choice of action when that action must be chosen before one can monitor the actions of those others". This definition emphasizes that the presence or absence of trust affects the decision of a truster, because actions of a trustee affect the well-being of the truster. The clause concerning the inability to monitor the trustee's actions is essential in analyzing trust formation. If an individual could monitor what a policy maker has done before she decides whether to trust the policy maker, the matter of trust would not arise.

In this section, a trust game is formulated in order to investigate trust formation between a policy maker and an individual regarding a water management plan. In this section, for simplicity we will refer to the policy maker and the individual as a trustee and a truster, respectively. The model we present modifies Bacharach and Gambetta's trust game (Bacharach and Gambetta, 2001). Bacharach and Gambetta built the theory of trust by formulating the trust game, while they studied an environment in which the structure of the game itself is of common knowledge to two players. Our model differs from that of Bacharach and Gambetta in that we remove the assumption of common knowledge and describe the trust game as a subjective game that represents the subjective frame of each player (Sugden, 1995; Bacharach and Stahl, 2000; Hatori and Kobayashi, 2011). In what follows, we first introduce the theory of the trust game developed by Bacharach and Gambetta (2001), and then propose a subjective trust game.

6.3.2 The Theory of Trust by Bacharach and Gambetta (2001)

Bacharach and Gambetta (2001) formulate a two-player strategic-form non-cooperative game, hereafter referred to as a trust game, in order to investigate trust formation between a truster and a trustee. Figure 6.2 shows the structure of Bacharach and Gambetta's trust game, while minor modifications are made. In the trust game, the truster has two actions labeled T and U, representing, respectively, "trust the trustee" and "distrust the trustee," respectively. On the other hand, the trustee's actions X and Y denote "make an effort for the truster" and "make no effort," respectively. In the context of water management, the truster's actions T and U represent her delegation of management authority to the trustee and vice versa. The trustee's action X means that a policy maker executes water management tasks properly, while action Y represents a moral hazard behavior.

Trustee

Truter		X	Y
	T	x, x	$\underline{x}, y$
	U	0, 0	0, 0

$x, y > 0,\ y > x,\ \underline{x} < 0$

Figure 6.2 Trust game.

The trust game aims to analyze "whether or not the truster trusts the trustee to choose strategy X", which, according to Bacharach and Gambetta, is *the primary trust problem*. In the trust game, the truster trusts the trustee if the truster expects that "the trustee chooses X." Furthermore, the trustee is said to be trustworthy if the next condition is satisfied; "if the trustee believes the truster trusts him, the trustee chooses strategy X". Bacharach and Gambetta call characteristics that make the trustee trustworthy *trust-warranting properties*. There are many possible trust-warranting properties, such as altruistic motivation, context-specific norms, rewards and punishment, to mention a few. The truster, however, rarely knows whether the trustee has trust-warranting properties. Bacharach and Gambetta suggest that the trustee's characteristics can be differentiated according to two properties, namely, *kripta* and *manifesta*. On the one hand, kripta denote unobservable internal properties of the trustee, like honesty, benevolence, time preference and so on. On the other hand, manifesta denote observable external properties including gestures, countenance and career.

Manifesta sometimes operate as signs of kripta and provide the truster with information regarding a trustee's internal characteristics. In general, trust-warranting properties of the trustee are unobservable kripta, and the truster cannot help drawing inferences as to the trust-warranting properties of the trustee from observation of trustee's manifesta. Sometimes, however, observable manifesta are less than fully reliable. Herein lies what Bacharach and Gambetta call *the secondary trust problem*, which focuses on the opportunistic behavior of a trustee who mimics trust-warranting properties through the display of manifesta. An opportunistic trustee who does not possess trust-warranting properties might deliberately display manifesta in order to be interpreted by the truster as a trustworthy type. For the analysis of this issue, Bacharach and Gambetta develop the secondary trust game allowing the trustee to provide the truster with certain information about his trustworthiness before the trust game. A signaling game in which the trustee and the truster are regarded, respectively, as sender and receiver, the secondary trust game deals with not only the question "Can the truster trust the trustee to choose strategy X?" but also the question "Are the manifesta that the trustee sends to the truster a reliable sign of trust-warranting properties?"

In what follows, we develop a model of trust formation based on the trust game by Bacharach and Gambetta, in which each player plays his or her own subjective game. As explained above, the difference between our model and Bacharach and Gambetta's is the assumption of the two players' common knowledge regarding the structure of the game. This context leads to important considerations absent in the analysis of the model by Bacharach and Gambetta. Especially, this approach enables us to analyze the effect on trust formation of the two players' differences in conceptual schemes.

6.3.3 The Subjective Trust Game

This study analyzes a situation where individual players draw different subjective games in their minds. Figure 6.3 shows the subjective framework of the trust game proposed in this study. The left side of this figure denotes the subjective game of the trustee, and the right side denotes that of the truster. This game differs from that of Bacharach and Gambetta in two ways. Firstly, for the same reason as that stated in the previous section, the model assumes that each player has a different language by which the player describes trustee's actions X and Y. Let the label that the trustee uses be called *action-label*, represented by a_X and a_Y. Similarly, let the label that the truster uses be called *outcome-label*, represented by o_X and o_Y. Secondly, each player maximizes subjective payoffs in his or her subjective game. Let us consider a situation where the truster holds stereotyped beliefs that the trustee never makes an effort for the truster, while in practice the trustee strives honestly to make an effort. Subjective payoffs described on the left and right sides of Figure 6.3 represent such a situation. We impose strong assumptions with regard to the truster's stereotyped beliefs in order to analyze in the following sections institutional designs under which the truster with stereotyped beliefs about a trustee's behavior comes to trust the trustee motivated to conduct management activities in accordance with the truster's will. The results that we derive from the analysis of the trust game in this section will serve as useful benchmarks when evaluating the institutions in subsequent sections.

Subjective game of Trustee

	a_X	a_Y
T	$x,\ x$	$\underline{x},\ \underline{x}$
U	$0,\ 0$	$0,\ 0$

Subjective game of Truster

	o_X	o_Y
T	$x,\ x$	$\underline{x},\ y$
U	$0,\ 0$	$0,\ 0$

$$x, y > 0,\ y > x,\ \underline{x} < 0$$

Figure 6.3 Subjective trust game.

The primary trust problem

First we examine trust formation in a situation where the two players make a choice without any communication between them. This situation corresponds to the primary trust problem according to the definition by Bacharach and Gambetta. Under the assumption of no communication, the problem as to differences in players' languages is moot. The analysis of the trust game aims to show whether the truster who has stereotyped beliefs about the trustee's behavior can trust the trustee.

In the trustee's subjective game, he always chooses action X. Then, consider the truster's choice in her subjective game. As a solution concept, we adopt Nash equilibrium. However, emphasis should be put on the fact that the solution is appropriate only in the truster's subjective game. This does not mean that the strategy profiles of the two players reach equilibrium in the truster's subjective game. The equilibrium strategy of the trustee in the truster's subjective game might differ from the actual trustee's choice in his subjective game, that is, action X. Let us assume that each player adopts mixed strategies. The strategy of the truster is represented by the probability τ that she chooses action T. In the same way, the strategy of the trustee is represented by the probability ρ that he chooses action X. In the subjective game, there exists a unique equilibrium ($\rho^* = 0$ and $\tau^* = 0$), where the truster chooses not to trust the trustee. In what follows, we call this equilibrium *distrust equilibrium*. The equilibrium strategy of the trustee, $\rho^* = 0$, represents the truster's stereotyped beliefs about the trustee's behavior. The result shows that the truster distrusts the trustee, although the trustee chooses action X in practice. Thus, unless the truster is able to understand that the trustee chooses action X, she does not trust him regardless of his actual choice.

The secondary trust problem

Next, for the secondary trust game, we consider a scenario where the two players communicate with each other before the basic trust game. The previous model is now modified to include pre-communication, in which the trustee uses a voice to gain trust from the truster. For example, we can consider a situation in which a policy maker evaluates his own management performance. The voice of the trustee is a statement about which action he will choose in the basic trust game. We impose an assumption that the strategy of manipulation incurs prohibitive expenses for the trustee and the trustee cannot send his voices strategically. This assumption means that the trustee is able to commit himself to choosing a certain action in advance. The truster who describes the trustee's actions by using outcome-labels, however, cannot understand the correct meaning of the trustee's voice represented by action-labels. As a result, the truster might choose to distrust the trustee. The purpose of the secondary trust game is to examine not only the question of "whether or not the truster trusts the trustee to choose the action X", but also the question of "whether or not the truster trusts the trustee's voice."

There exist the following one-to-one mappings as possible correspondences between action-labels and outcome-labels:

$$\varphi_t(a_X) = o_X, \varphi_t(a_Y) = o_Y$$
$$\varphi_f(a_X) = o_Y, \varphi_f(a_Y) = o_X$$

where the mapping φ_t denotes the true correspondence between the two labels, while the mapping φ_f denotes the false correspondence. The truster is uncertain about the true correspondence, and she has a subjective probability β that she assigns to the event that mapping φ_t is true. The secondary trust game assumes that subjective beliefs are not updated through a process of communication, because the truster cannot draw any inferences about the true correspondence from the communication. This assumption will be removed in section 5, where the truster is able to receive information about the correspondence.

In the subjective game of the trustee, he chooses action X in the same way as in the basic trust game and sends voice a_X to the truster. On the other hand, the truster's action in her subjective game depends upon her subjective beliefs β. There exist two equilibria: 1) equilibrium (hereafter *trust equilibrium*), where the truster trusts the trustee, and 2) distrust equilibrium, where the truster does not trust the trustee.

1) Trust equilibrium

If the subjective beliefs β satisfy the condition $\beta \geq \beta^*$, the truster trusts the trustee ($\tau^* = 1$).

2) Distrust equilibrium

If the condition $\beta < \beta^*$ is satisfied, the truster does not trust the trustee ($\tau^* = 0$).

β^* in the above condition is denoted by

$$\beta^* = \frac{-\underline{x}}{x - \underline{x}}.$$

For trust equilibrium, the condition $\beta \geq \beta^*$ is required. More precisely, it is required that the truster perceive the trustee's action correctly. On the other hand, if the truster anticipates that the trustee will choose the action represented by o_Y, she distrusts the trustee. Thus, whether or not the truster trusts the trustee depends upon the subjective beliefs of the truster. In a case where the truster misunderstands trustee's voice, the trust equilibrium cannot be realized. Indeed, it is difficult for people who lack knowledge about the technical conditions of regional management to understand what the policy maker says and does. As a result, they might not trust the policy maker.

6.4 LANGUAGE LEARNING AND TRUST FORMATION

This section considers that both language learning and trust, embodied in the range of CWM, are achieved through experience. First, common language can be attained through learning by doing in the regional practice of CWM. Players can gain information about the logic model discussed in section 6.2 from such regional experience. Second, according to Luhman (1979), it is also possible to view trust as being conceived in terms of social practices. As indicated previously, Luhman characterizes trust as a mechanism that reduces complexity and enables people to cope with the high levels of uncertainty and complexity of contemporary life. According to him, trust is a matter of induction rather than deduction in that it "extrapolates from the available evidence". Trust relations are produced and reproduced through daily interactions. Thus, both regional practices can be seen to be based on forward induction. Viewed in this way, they are close in spirit to an "*unjustified leap in the dark*" (Kripke, 1982).

In order to examine interrelated learning and trust formation, we develop a simple model of the interplays between a policy maker and an individual (median voter) in the process of regional decision-making for water management. Our analysis shows that the policy maker might face a trade-off between learning an optimal solution and gaining the trust of an individual; learning by trial and error might interfere with trust formation, and vice versa. Accordingly, the optimal project might not be carried out.

6.4.1 The Dynamic Model

We analyze a repeated two-player game in which a policy maker and an individual are the players, but they lack a common description of the game. The stage game is comprised of the subjective communication game and the subjective trust game developed in sections 6.2 and 6.3, respectively. Here, as in the previous model, the policy maker must make a decision about whether or not to undertake a project in regional management and if so what project to carry out. If the policy maker decides not to undertake any project, the status quo is maintained. The outcomes of regional decision in each period depend on the individual's well-being, which is not known in advance to the policy maker. They can be observed after being selected in the stage game. Observing the outcome in the stage game, the individual decides whether to trust or distrust the policy maker. The model assumes the possibility of moral hazard behavior of the policy maker, as in the trust game. The stage game can be repeated in periods $t = 1, 2, \ldots$ until the individual decides to distrust the policy maker.

Private description of projects

In the same way as in section 6.2, alternatives are represented by two different languages: action labels for the policy maker, denoted by $\boldsymbol{A} = (a_0, a_1, \ldots, a_n)$, and outcome labels for the individual, denoted by $\boldsymbol{O} = (o_0, o_1, \ldots, o_m)$. Players distinguish alternatives roughly by "success" or "failure". For simplicity, let an ideal action a_d and an ideal outcome o_d be "success" and others be "failure". Any assignment of a success across alternatives is equally likely.

Learning from failure

As in previous research, the lack of common description can be expressed in terms of symmetry restrictions on players' strategies (Blume, 2000; Crawford & Haller, 1990). We assume that the policy maker uses strategies invariant with the players' private descriptions, namely *attainable strategies*. Attainability requires the policy maker to put equal weight on all the actions that he has not used before.

The model requires players to learn from failures. If a success has not been achieved at the previous stage, this observation serves as a common reference for the next stage. This means that the policy maker revises his beliefs about project outcomes and avoids projects that have led to failures in the past. Thus, the probability of a success in period t is represented by $\alpha_t = 1/n_t$, where n_t is the number of possible projects in that period.

Trust formation

After observing the management outcome in each period, the individual decides whether or not to entrust the next regional management project to the policy maker. If the individual trusts the policy maker, the policy maker retains decision authority in the next period. If the individual distrusts him, however, the policy maker is deprived of decision authority, and the game ends. We suppose that the status quo is maintained after the end of the game. The individual possesses stereotyped beliefs that the policy maker possibly makes no effort on behalf of the individual, while in practice the policy maker makes an honest effort. The moral hazard behavior of the policy maker is denoted as a_M (o_M), which corresponds to a failure in outcome label. The subjective probability in period t that the individual assigns to the event that the policy maker is honest is represented by π_t.

Finally, the payoffs of both players in the stage game depend on the outcomes of selected alternatives. At a success, all players obtain a payoff $x > 0$. At a failure, they get $\underline{x} < 0$, while the individual believes that the policy maker earns $x > 0$ even in a case where he is dishonest and makes no effort. Regarding the expected payoffs from a project, the following assumption is satisfied:

Assumption 1. $\alpha_0 x + (1-\alpha_0)\underline{x} > 0.$

As $\alpha_t \geq \alpha_0$ for every t, the above condition implies that learning by doing is always an optimal strategy for both players.

6.4.2 Equilibrium Play

As a solution of the present model, we employ a *full equilibrium* proposed by McKelvey and Ordeshook (1985). In this equilibrium, each player chooses a strategy that maximizes his payoffs subject to his beliefs. The policy maker randomizes over actions equally based on his belief α_t in period t. The individual trusts the policy maker when current management results in a success. At a failure, the individual's equilibrium strategy is represented by

$$\begin{cases} \text{Trust: if } \pi_t \alpha_t x + \{\pi_t (1-\alpha_t) + (1-\pi_t)\} \underline{x} \geq 0 \\ \text{Distrust: otherwise.} \end{cases} \tag{6.2}$$

The above condition means that the expected payoffs the individual can gain at the next period if she entrusts the next decision to the policy maker are larger than the payoffs of the status quo. Let π^* be the marginal belief that satisfies the equality of (6.2). Furthermore, full equilibrium allows for revision of beliefs based on the observed strategy choices of other players. We assume that after observing failure, the individual updates her belief π_t by using Bayes' Rule:

$$\pi_t = \frac{\pi_{t-1}\alpha_t}{\pi_{t-1}\alpha_t + (1-\pi_{t-1})} \tag{6.3}$$

The equilibrium strategy of the individual can be summed up in the following lemma:

Lemma 1: *The individual trusts the policy maker if and only if a success is played or* $\pi_t > \pi^*$. *Otherwise, the individual distrusts the policy maker.*

From equation (6.3), it can be seen that the individual's beliefs satisfied the condition $\pi_t < \pi_{t-1}$. This implies that the policy maker loses the trust of the individual as the game proceeds.

Then, there exists a period t^* such that $\pi_{t^*} < \pi^* < \pi_{t^*-1}$. We have thus established the following proposition:

Proposition 1: *In the dynamic model of the repeated two-player game, the individual trusts the policy maker if $t \leq t^*$. Otherwise, the individual distrusts the policy maker.*

6.5 POLICY GAME OF THIRD PARTY REVIEWS

Proposition 1 shows that players are unable to sustain a truster-trustee relationship insofar as they learn by trial and error only through observing project outcomes. Learning in this way can destroy relationships built on trust, and the absence of trust discourages further learning.

In this section, we focus on the role of third party reviews to support the trust formation process between the policy maker and the individual. We investigate trust formation between two players in a situation where the policy maker's action is evaluated by one or more third party reviewers. Thus the individual can obtain information about water management activities from the report of a knowledgeable third party.

The use of third party reviews in the determination of regional decision-making has become increasingly widespread. Indeed, local governments often hold public forums where various experts and interested parties deliberate the desirability of a regional project. Our motivation in introducing third party reviews is to give the individual an objective reason to trust the policy maker.

6.5.1 Assumptions

To understand the institution of third party reviews, we extend the previous dynamic model by adding a third party who makes announcements regarding the actions of the policy maker. The reviewer is assumed to be a water expert or an interested party related to watershed decision-making. It is assumed that third party reviewers have the ability to monitor the management activities of the policy maker correctly, in that they possess perfect information about the policy maker's actions. After observing such actions, reviewers send to the individual their voices about which action the policy maker chooses. Reviewers, however, might have private interests in the decision-making problem. Furthermore, the policy game allows reviewers to send their voices strategically to the individual. In this context, reviewers are likely to express their voices not from a neutral standpoint but from a standpoint reflecting a specific interest. Accordingly, they are able to manipulate a policy maker's action. Let us define the reviewer's voice as the "correct voice" if the voice corresponds to a policy maker's actual action, and a "wrong voice" otherwise. In addition, reviewers might conceive a policy maker's actions in terms of their own language, just as the policy maker and the individual do. It is assumed that reviewers are differentiated into two types depending upon their interests and languages: 1) a reviewer who prefers that the individual trust the policy maker and who possesses the same language as the policy maker, i.e. action-labels (this type of reviewer is hereafter called "*reviewer P*"), and 2) a reviewer who prefers that the individual distrust the policy maker and who possesses the same language as the individual, i.e. outcome-labels (hereafter called "*reviewer Q*"). In section 6.5.2, we show that the mechanism of checks and balances among the reviewers cannot be worked well because of differences in language. In section 6.5.3, we introduce an interpreter as an additional reviewer who bridges the gulf between other reviewers' opinions caused by differing languages.

6.5.2 Third Party Reviews With Different Languages

In the policy game, the payoffs of reviewers are determined by their types and the results of the stage game formulated in the previous section. Reviewer P obtains a positive payoff normalized to unity if the individual trusts the policy maker in the stage game, while the reviewer obtains payoff zero otherwise. The reverse holds for the payoffs of reviewer Q; that is, the payoff of reviewer Q is unity if the individual chooses not to trust the policy maker, and zero otherwise. As explained in the previous subsection, reviewer P and reviewer Q each have their own language by which they describe the action of the policy maker. Thus, reviewer P sends voice $m^P \in (a_0, a_1, ..., a_n)$, while reviewer Q sends voice $m^Q \in (o_0, o_1, ..., o_n)$. It is assumed that there is mutual knowledge among all players about the types of third reviewers. The strategies of the policy maker and the individual are denoted by ρ and τ, respectively, in the same way as in the trust game. The strategy of the third reviewer is denoted by $\boldsymbol{\sigma}^r = (\sigma_0^r, \sigma_1^r, \dots, \sigma_n^r)(\mathrm{r} = \mathrm{P}, \mathrm{Q})$, where σ_i^r $(i=0, \dots, n)$ denotes the probability that reviewer P (Q) sends voice a_i (o_i). Since the choice of the third reviewer is conditioned on action i of the policy maker $(i = 0, \dots, n)$, we denote the strategies of the two reviewers as $\boldsymbol{\sigma}^P(a_i)$ and $\boldsymbol{\sigma}^Q(o_i)$. The choice of the individual depends on the voices of the reviewers. Therefore, the strategy of the individual is represented by τ (m^P, m^Q).

The policy game assumes a peer review system comprised of two types of reviewers. The following analysis aims to examine the mechanism of checks and balances between reviewers. By having one reviewer's statement crosschecked with the other reviewer's, this mechanism forces reviewers to be truthful. With such checks and balances, both reviewers cannot but send a correct voice about a policy maker's action. We assume that in the peer review system each reviewer can obtain an additional positive payoff $\mu > 0$ if the voices of the two reviewers coincide with each other, more precisely, $\varphi_t(m^P) = m^Q$ (Frisell, 2002). The payoffs μ are referred to as statement rewards hereafter. Statement rewards represent both pecuniary rewards that a reviewer is paid for the results of evaluation and non-pecuniary rewards such as prestige and reputation. Let us assume that the result of the stage game is more important for each reviewer than statement rewards, i.e. $1 > \mu > 0.5$.

Apparently, trust equilibrium is more likely to be generated with multiple reviewers, since the individual can obtain more information about the action of the policy maker. As will be shown, however, a difference in languages of the two reviewers invalidates the mechanism of checks and balances and leads to distrust equilibrium. To understand the intuition behind this result, let us assume a situation where the mechanism of checks and balances works well. In this context, the strategy of each reviewer is represented by

$$\sigma_i^P(a_i) = 1, \ \sigma_i^Q(o_i) = 1 \tag{6.4}$$

Since reviewers send the correct voice to the individual, the strategy of the individual is

$$T\left(m^P, m^Q\right) = \begin{cases} 1: if\ m^Q = o_{-M} \\ 0: if\ m^Q = o_M \end{cases} \tag{6.5}$$

The above strategy means that the individual depends on the voice of reviewer Q, who shares the same language with the individual. If reviewer Q follows the strategy (6.4), the individual can get correct information about the action of the policy maker. Accordingly, the policy maker makes an effort in the individual's subjective game, because the policy maker cannot gain trust from the individual if he deviates from the strategy. These strategies, however, are not robust against players' deviations. The strategy (6.5) of the individual is decided only according to the voice of reviewer Q. This is because the individual cannot understand the voices of reviewer P, which are denoted by action-labels. However, this strategy gives reviewer Q an incentive to deviate from sending the correct voice, a deviation which, in turn, makes the individual disregard the voice of the reviewer. As a result, the individual who has no information about the action of the policy maker does not trust him. Accordingly, distrust equilibrium is obtained as a unique equilibrium of this policy game. Since reviewers cannot affect the behavior of the individual, they are indifferent to which voices to send. Equilibrium strategies of reviewers are represented by $\boldsymbol{\rho}^{r*} = con$ (r = P, Q), where *con* denotes the constant value given independently of the action of the policy maker.

As a result, the individual updates her belief π_t by using Bayes' Rule (3) in the same way as in the previous model. Therefore, also in this policy game, the policy maker loses the trust of the individual as the game proceeds. We have thus established the following proposition:

Proposition 2: *In the policy game of third party reviews with different languages, the individual trusts the policy maker if $t \leq t^*$. Otherwise, the individual distrusts the policy maker.*

6.5.3 Third Party Reviews with a Common Language

In the former policy game, the individual cannot judge whether or not the voice of one reviewer is correct by comparing it with the voice of the other reviewer, since the two voices are represented by different languages. As a result, the individual remains skeptical of reviewers' voices. For the mechanism of checks and balances to work well, it is mandatory that the correspondence between reviewers' voices be specified, and that all players share an understanding of the different languages. In this subsection, we introduce an additional reviewer to verify the correspondence between the two languages. Let us call the additional reviewer an *interpreter*. It is assumed that the interpreter knows the true correspondence φ_t and sends a voice about the correspondence to the individual. The voice of the interpreter is given by $m^I \in \Phi$. In what follows, the correspondence that the interpreter proposes is defined as the *logic model*. Let us consider that the interpreter also has preferences regarding the result of the basic trust game. We assume that the interpreter is of type P, while the same result is obtained even if the interpreter has the same motivation as reviewer Q. Let us consider that reviewers' voices (m^P, m^C, m^I) are consistent with each other if the voices satisfy the condition $\varphi_t(a_i) = o_i$. We assume that three reviewers involving the interpreter derive statement rewards μ if their voices satisfy the consistency condition.

The strategies of the policy maker, the individual and reviewers are set by $\tau, \rho, \boldsymbol{\sigma}^P$ and $\boldsymbol{\sigma}^Q$, respectively. The strategy of an interpreter is denoted by $\boldsymbol{\omega} = (\omega_1,..., \omega_R)$, where ω_r represents the probability that the interpreter sends voice $\varphi_r \in \Phi$. In this model, the individual receives voices from two reviewers and the interpreter. Accordingly, the individual's strategy is denoted by τ (m^P, m^C, m^I). This policy game has trust equilibrium where the individual trusts the policy maker according to the voices of three reviewers throughout the game. In equilibrium, the individual chooses the following strategy:

$\tau^*(m^P, o_{-M}, m^I) = 1$: if messages (m^P, o_{-M}, m^I) are consistent.
$\tau^*(m^P, o_M, m^I) = 0$: if messages (m^P, o_M, m^I) are consistent.
$\tau^*(m^P, m^C, m^I) = 1/2$: if other messages are sent.

Thus, the individual chooses to trust the policy maker, if and only if the voices of reviewers are consistent with each other and both reviewers make an announcement that the policy maker has made or is making an effort. On the other hand, she does not trust the policy maker if both reviewers inform her that the policy maker is choosing or has chosen moral hazard behavior. In cases where reviewers' voices do not satisfy the consistency condition, the individual chooses mixed strategies, because she cannot judge which voice is correct. As for the third reviewers, they send the correct voice (strategy (6.4)). Similarly, an interpreter informs the individual about the true correspondence ($\omega_t^* = 1$). Receiving the voices of reviewers who follow the above strategies, the individual can gain perfect information about the action of the policy maker ($\pi_t = 1; t = 1, 2, \ldots$) and thus believe that the policy maker is honest.

In trust equilibrium, the mechanism of checks and balances is worked well, and the voice of one reviewer is always checked by the voice of the other reviewer. The mechanism of checks and balances is sustained by the logic model that the interpreter presents. Since different languages are connected clearly by the logic model, the individual can understand the meaning of reviewers' voices. As a result, any deviation from the equilibrium is worse for both reviewers, and accordingly each reviewer is forced to be truthful.

Proposition 3: *In the policy game of third party reviews with common languages, the individual trusts the policy maker throughout the game.*

6.6 CONCLUSION

This chapter focused on two conditions upon which the potential of interrelated water management systems depends, i.e. common language and trust. The theoretical model showed that differences in cognitive schemes of a policy maker and an individual (median voter) prevent them from communicating with each other, and then reduce the likelihood of trust formation. The analysis of the model indicated the importance of attaining shared descriptions by formulating a logic model regarding the correspondence between two different languages. Our analysis also showed that the policy maker might face a trade-off between learning an optimal solution and gaining the trust of the individual, and as a result the optimal project for water management might not be carried out. We discussed ways to achieve language learning and trust formation simultaneously.

Finally, we will discuss the implications of our analysis on CWM.

The roles of reviewers and interpreters

The result of the theoretical analysis in this chapter suggests that it is necessary to set up a forum organized by multiple third reviewers who share common languages individually with a policy maker and an individual, and an interpreter who connects different languages. Especially clear from our analysis is that the role the interpreter plays in checking differences in reviewers' voices is indispensable for trust formation in CWM.

The kind of labels each player uses depends on his or her knowledge, experience, psychological and cultural factors, and so on. Therefore, the interpreter who formulates the logic model should be sufficiently talented as a professional authority possessing knowledge regarding both the technical conditions of an alternative (*action-labels*) and the influences the alternative has on residents' life (*outcome-labels*).

Sufficiently bad memory and/or minds of tolerance

The dynamic model assumes that players have perfect memory of past events. The result of the dynamic model shown in section 6.4 seems to suggest that such perfect memory discourages sustainable learning for CWM. In the model, the individual judges whether or not to trust the policy maker based on information about all previous events. This might enable her to make a deliberate decision, but on the other hand it might reduce her tolerance with regard to the policy maker, thereby decreasing the individual's flexibility to adapt to the current problem. Thus, we can say that regional sustainability of learning activities demands sufficient bad memory and/or minds of tolerance.

Learning from success

The dynamic model assumes the process of learning from failures. Indeed, to experience a failure can give people important lessons useful to successful regional management in the future. But to experience only failure might decrease the learning potential of the region. Our analysis shows that the failure experience decreases regional members' trust in policy makers. Thus, while failure can be useful to the learning process, success stories are vital to fostering trust among those who manage water resources. It is important that policy makers gather examples of success in CWM and share them with regional actors.

6.7 REFERENCES

Bacharach M. and Stahl D. (2000). Variable-frame level-n theory. *Games and Economic Behavior*, **32**(2), 220–246.

Bacharach M. and Gambetta D. (2001). Trust as type detection. In: Deception, Fraud and Trust in Virtual Societies, C. Castelfranchi and Y.-H. Tan (eds.), Kluwer Academic Publishing, Dordrecht, the Netherlands, pp. 1–26.

Barber B. (1983). *The Logic and Limit of Trust*. Rutgers University Press, New Brunswick, NJ.

Biswas A. K. (2004). Integrated water resources management: A reassessment. *Water International*, **29**(2), 248–256.

Blume A. (2000). Coordination and learning with a partial language. *Journal of Economic Theory*, **95**(1), 1–36.

Camerer C. (2003). Behavioral studies of strategic thinking in games. *TRENDS in Cognitive Sciences*, **7**(5), 225–231.

Casajus A. (2000). Focal points in framed strategic forms. *Games and Economic Behavior*, **32**(2), 263–291.

Crawford V. and Haller H. (1990). Learning how to cooperate: Optimal play in repeated coordination games. *Econometrica*, **58**(3), 571–595.

Crawford V. and Sobel J. (1982). Strategic information transmission. *Econometrica*, **50**(6), 1431–1451.

Cho I.-K. and Kreps D. (1987). Signaling games and stable equilibria. *Quarterly Journal of Economics*, **102**(2), 179–221.

Dasgupta P. (1988). Trust as a commodity. In: Trust: Making and Breaking Cooperative Relations, D. Gambetta (ed.), Basil Blackwell, Oxford.

Devetaga G. and Warglien M. (2008). Playing the wrong game: An experimental analysis of relational complexity and strategic misrepresentation. *Games and Economic Behavior*, **62**(2), 364–382.

Farrell J. (1993). Meaning and credibility in cheap-talk games. *Games and Economic Behavior*, **5**(4), 514–531.

Ferreyra C., de Loë R. C. and Kreutzwiser R. D. (2008). Imagined communities, contested watersheds: Challenges to integrated water resources management in agricultural areas. *Journal of Rural Studies*, **24**, 304–321.
Festinger L. (1957). *A Theory of Cognitive Dissonance*, Row, Peterson, Evanston, IL.
Fox C.R. and Rottenstreich Y. (2003). Partition priming in judgment under uncertainty. *Psychological Science*, **14**(3), 195–200.
Frisell L. (2002). Taking advice from imperfectly informed lobbyists: When to match hawks with hawks. Mimeo, the Riksbank, Stockholm.
Hatori T. and Kobayashi K. (2011). Public reviews and trust formation in disaster risk management. *Journal of Natural Disaster Science*, **32**(2), 89–97.
Hering J. G. and Ingold K. M. (2012). Water resources management: What should be integrated? *Science*, **336**, 1234–1235.
Kahneman D. and Tversky A. (1979). Prospect theory: An analysis of decision under risk. *Econometrica*, **47**(2), 263–291.
Kahneman D. and Tversky A. (eds.), (1984). Choices, values, and frames. *American Psychologist*, **39**(4), 341–350.
Kobayashi K. and Hatori T. (2006). Forest governance: Structure and function. In: Social Capital and Development Trends in Rural Areas, **Vol. 2**, K. Ito, H. Westlund, K. Kobayashi and T. Hatori (eds.), MARG, pp. 23–40.
Kreps D. (1990). *Game Theory and Economic Modeling*, Oxford University Press, Oxford.
Kripke S. (1982). *Wittgenstein on Rules and Private Language*, Harvard University Press, Cambridge, Mass.
Luhmann N. (1979). *Trust and Power*, Wiley, Chichester, U.K.
Lupia A. and McCubbins M. D. (1998). *The Democratic Dilemma; Can Citizens Learn What They Need To Know?* Cambridge University Press, Cambridge, UK.
Margerum R. D. (1999). Integrated environmental management: The elements critical to success. *Environmental Management*, **24**(2), 151–166.
Mckelvey R. D. and Ordeshook P. C. (1985). Elections with limited information: A fullfilled expectations model using contemporaneous poll and endorsement data as information sources. *Journal of Economic Theory*, **36**(1), 55–85.
McLaughlin J. A. and Jordan G. B. (1999). Logic models: A tool for telling your programs performance story. *Evaluation and Program Planning*, **22**(1), 65–72.
Piñon A and Gambara H. (2005). A meta-analytic review of framing effect: Risky, attribute and goal framing. *Psicothema*, **17**(2), 325–331.
Rabin M. (1990). Communication between rational agents. *Journal of Economic Theory*, **51**(1), 144–170.
Robert J. and Scapens R. (1985). Accounting systems and systems of accountability: Understanding accounting practices in their organizational contexts. *Accounting Organizations and Society*, **10**(4), 443–456.
Rubinstein A. (1998). *Modeling Bounded Rationality*, MIT Press, Cambridge, MA.
Selten R. (1990). Bounded rationality. *Journal of Institutional and Theoretical Economics*, **146**(4), 649–658.
Simon H. (1995). Rationality in political behavior. *Political Psychology*, **16**(1), 45–61.
Sneddon C. (2002). Water conflicts and river basins: The contradictions of comanagement and scale in Northeast Thailand. *Society and Natural Resources*, **15**(8), 725–741.
Spector M. B. and Kitsuse J. I. (1977). *Constructing Social Problems*. Cummings, Menlo Park, CA.
Sugden R. (1995). A theory of focal points. *The Economic Journal*, **105**(430), 533–550.
Van Bueren E. M., Klijn E-H. and Koppenjan J. F. M. (2003). Dealing with wicked problems in networks: Analyzing an environmental debate from a network perspective. *Journal of Public Administration Research and Theory*, **13**(2), 193–212.
Wittgenstein L. (1957). *Philosophical Investigations*. Blackwell, Oxford.

7 The right to water and the participatory approach

Hayeong Jeong

Graduate School of Management, Kyoto University, Yoshida Hon-cho 36-1, Sakyoku, Kyoto, 606-8317, Japan. E-mail: jeong.hayeong.2z@kyoto-u.ac.jp

Abstract

Water scarcity is one cause of exclusion from access to water. Everyone has a right to access to water as a primary resource for living. However, current markets and government systems are unable to ensure this basic right. To address this problem, right-based water management is discussed. In this study, first the cause of exclusion due to water scarcity is described in terms of "accessibility and availability." Next, how the participatory approach to water management involving right reallocation can improve the water rights of vulnerable users is discussed by reviewing previous water rights and collective actions by community members. Finally, the limitations of the participatory approach are discussed.

Keywords: Community-based water management, Exclusion, Includable right, Water Scarcity.

7.1 INTRODUCTION

Water is a primary resource for everyday living. In pre-industrial society, water was viewed as a non-tradable, disposable, and abundant commodity. Present-day population growth and large-scale production require a large amount of water. Over-exploitation of water leads to the degradation of its quantity and quality, which in turn leads to severe water scarcity, often resulting in conflict and the exclusion of vulnerable water users.

Even if a government provides public water service, it is difficult to avoid conflict and exclusion problems. Marginal rural and suburban areas are easily dismissible from public water service because of the limited development of infrastructure such as roads and electricity. In such marginal areas, water is a resource associated with exclusion and rivalry. According to UNICEF (2007), about 655 million people, i.e., 17.6 percent of the world, lack access to safe water; about 1.8 million children die annually; and about 433 million school days are missed due to diarrhea from unsafe water and poor sanitation. Most of those affected are living in marginal areas.

The potential impact of water scarcity on vulnerable water users is tremendous. Water scarcity negatively affects not only human health and the ecosystem but also economic and social development. Gasper and Truong (2008) pointed out the importance of development ethics and introduced to development the hot issues of the international human rights system and right-based approaches (i.e., the moral right to life). The idea of right-based approaches is also emphasized to manage water resources for 'efficiency and equity of water right allocation'. In November 2002, the UN Committee on Economic, Social and Cultural Rights (CESCR) adopted a General Comment on the Right to Water. For the first time, water is explicitly recognized as a fundamental human right, and the 145 countries which have ratified the international CESCR will now be compelled progressively to ensure that everyone has access to safe and secure drinking water, equitably without discrimination. Since water has come to be perceived from the perspective of scarcity, it has become associated with human rights, as water is necessary for sustaining the life and health of humans. The Office of the High Commissioner for Human Rights (OHCHR) defined the elements of a rights-based approach to development as stated below.

- Express linkage to rights
- Accountability
- Empowerment
- Participation
- Non-discrimination and attention to vulnerable groups

How can water rights be secured for vulnerable water users? We advocate a new development orthodoxy advancing economic and social human rights by involving participatory democracy that promotes social inclusion and self-determination in the water management process, including decision-making, planning, and allocation. Cowen and Shenton (1998, p.55) stated, "Development is construed as 'a process of enlarging people's choices'; of enhancing 'participatory democratic processes' and the 'ability of people to have a say in the decisions that shape their lives'; of providing 'human beings with the opportunity to develop their fullest potential'; of enabling the poor, women, and 'free independent peasants' to organize for themselves and work together." The goal of development is to promote not only economic growth but also equity and self-reliance. A community-based management system (hereafter referred to as "CBMS"), the practice of a participatory approach to water management by vulnerable water users utilizing their social capital, is a reliable and sustainable way to ensure such users' water right. This study discusses the validity of the participatory approach to water management by vulnerable water users.

7.2 BASIC IDEA

7.2.1 Water Scarcity and Exclusion

The demand for usable water is determined by a variety of factors, including population growth, technical advancements, allocation mechanisms such as trade agreements, and environmental requirements (Zabihollah, 1990). Usable water is dramatically decreasing because of not only a 'water shortage' caused by overexploitation and the deterioration of water quality due to contamination stemming from socio-economic activities growth, but also the 'failure of water allocation' in terms of a 'high degree of subtractability and excludability' due to privatization and conservation policies.

The causes of water scarcity can be distinguished into two types: 'water availability' and 'water accessibility.' Water availability refers to the renewable capacity of water, which has seasonal, regional and technical variations. Recent advancement in the world's technology for storage, purification, and transport utilities has improved water availability. On the other hand, water accessibility refers to the right to use water, which has contextual and systemic variations. Normative structures of social interactions determine water accessibility. Asia and the Pacific areas are rich in renewable water resources, but water accessibility is the second lowest in the world, mainly due to population growth and undeveloped infrastructure (ESCAP, 2009).

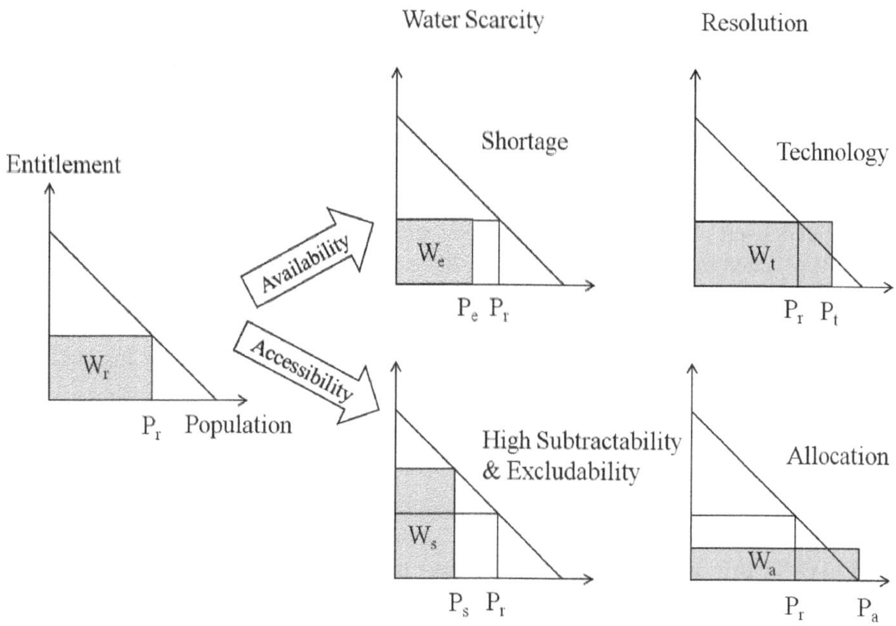

Figure 7.1 Water availability and accessibility.

Figure 7.1 describes two causes of water scarcity: availability and accessibility. Suppose that there is a certain amount of water W_r for number of population P_r with a basic level of entitlement[1]. If water scarcity generated by availability, such as a water shortage, occurs, then the number of people who can access the water falls to the level of P_e. The number of r – e population does not have access to water. Drought and depletion are representative examples of water scarcity generated by availability. A solution to water scarcity generated by availability is to improve the technology for storing or reusing water. On the other hand, if water scarcity generated by accessibility in terms of a 'high degree of subtractability[2] and excludability' by misbalancing the level of entitlement W_s, is biased toward a certain group, then the number of r – s people with access to water will be decreased to P_s. People with weak entitlement are preferentially excluded (Sen, 1981 and 1999, Sato, 2002). Representative examples are monopoly pricing/over-pricing of water and the water conservation policy of the government. A resolution of water scarcity generated by accessibility is to find an allocation system.

---

[1]An entitlement is a guarantee of access to a resource based on established rights or by legislation, which generally strengthens ownership.

[2]Subtractability refers to the degree to which one person's use of a resource diminishes others' use.

Technology improvement and allocation system are keys to reduce water scarcity. Even if there is technology improvement, vulnerable people will be excluded without any allocation system. In the same way, even if there is an allocation system, vulnerable people will be excluded without technology improvement. Institutional coordination to simultaneously improve technology and allocation system is necessary.

Nakastuka *et al.* (2004) simulated water availability for six continents: Africa, Asia, Australia, Europe, North America, and South America. The dynamics for analyzing the world water simulation model is based on system dynamics for analyzing hierarchic feedback structures and dynamic characteristics. The model considered seven factors: population, capital, agriculture, nonrenewable resources, persistent population, water quantity and water quality. The simulation result predicted a deficit of water in Asia from 2010 and North America from 2030, due to an unbalance between supply and demand. In addition, Molle (2004) pointed out the problem of exclusion of 'small users' in the rights allocation process. Small users are granted access to water without a permit and typically include people withdrawing water for domestic use, backyard gardens or livelihood use, in contrast with bulk users (such as urban water utilities, large irrigation scales and industries) that are granted an official license (or permit). Molle said that the poor especially are generally unaware of their rights, are unfamiliar with administrative/legal processes and have an instinctive (and understandable) reluctance to engage in them.

7.2.2 The Poor and Water Rights

Many poor people highly depend on local resources for their living (Durning, 1989). Sen (1981) defined the poor as those who cannot consume the crops they produce because they do not have land property, cannot change their property with food due to excess supply, involuntarily lost their jobs due to a demand decrease, and have lost an important source of income. As a result, a mouthful of any remaining food flows from starving people to those who are not starving, and from a region in which the minimum requirement is not met to a relatively rich region. People or regions with few or no exchangeable resources succumb in a competition with people with strong entitlement.

Entitlement competition limits the right of access to water (Sen, 1981; Dasgupta, 1993; Sato, 2000). Capitalistic society concentrates individual economic power in private property, which is a key factor in entitlement allocation. Consequently, poor people who lack private property will always lose the right of access to water. Traditionally, water was treated as a common-pool resource and managed by local residents living close to water springs. As the entitlement system strengthens, with private ownership bestowing the right to use water and with the economic value of the water resource increasing, local residents' access to water is restricted. If poor people are restricted from accessing resources because of deterioration or scarcity, they might be narrow-minded and resort to taking resources in unplanned and illegal ways. It is not easy for poor people to change the situation because their economical and political signals are insufficient to allow them to strengthen their rights. With resource allocation based on entitlement competition such as through the market mechanism, poor people might remain without access to water and in extreme cases be exiled from their community altogether.

As an alternate solution to entitlement competition, we focus on the capability approach (Sen, 1999) and willingness to pay (Hirschman, 1970). Sen (1999) said that people who were born in repressive societies, such as slaves or laborers with no ways or means to raise revenue except by their labor, are excluded not only from the viewpoint of welfare, but also from the viewpoint of "corresponding" basic capabilities, e.g. the ability to be well-nourished and well-sheltered, the capability to escape avoidable morbidity and premature mortality, and so forth (capability approach). Corresponding basic capabilities depend on basic liberty. Sen stated that responsibility requires liberty as a precondition. Unlike Sen (1999), Hirschman (1970) argued that the institutional system ensures rooting added value derived from the scarcity in the local area and using the resource. One system is to enhance residents' willingness to pay for a regional resource. When the willingness to pay (i.e., resource loyalty) is enhanced, residents spend more time on conserving resources, and they become more responsible for resources. Resource rights based on the willingness to pay would be an effective solution. However, it is not easy to enhance poor people's willingness to pay for the development and conservation of regional resources. In order to enhance the resource loyalty of the people, it is necessary to transfer the long-term management right of the resource to local residents and improve their daily life conditions in the region. In addition, the evaluation methodology to determine the corresponding basic capability and the willingness to pay should be discussed.

7.2.3 The Resource Allocation Mechanism

In a capitalistic society, resource allocation has occurred through two mechanisms, 'market' and 'government' (Sato, 2002). Resource allocation by market accomplishes Pareto efficiency[3] (AM, 1999). It is based on private ownership, which turns into market trade and a marketing foundation generating the commercialization of resources (Simon, 1996). On the other hand, resource allocation by government accomplishes simplification in order to easily monitor and control resources and to normalize resources based on the units and rules decided by the government (Sato, 2002). Resource scarcity has the role of enforcing not only the government, which wants to reserve the resource, but also the market, which wants to create profits and reacts to price (Simon, 1996).

Utilitarianism insists on the idea that society should work toward the greatest happiness for the greatest number. That is, resource allocation should be for maximizing the total utility (satisfaction) of all society members. However, the problem with this thinking is that people are treated as machines which create utility by investing in the resource. In contrast, Rawls (1971) suggested Max-min equity principle[4] of social primary goods as an allocation principle, which avoids the problem of comparing the utility difference between individuals and the problem of utilitarianism. Social primary goods means coveted goods for people who are rational, such as rights, liberty, chance, income, and financial resources. Also, Siono (1984) said that the use of resources should depend on decisions about the resource: how the resource is allocated, what values are considered by whom, and finally who decides the allocation. Resources are not only the goods themselves but also things used for purposes based on the evaluation of

[3]This is a state of allocation of resources in which it is impossible to make any one individual better off without making at least one individual worse off.

[4]The Maxmin principle is a decision rule to max people who have minimum social primary good. Dworkin (1981) and Roemer (1986) argued that equality of resources should create equality of welfare, e.g. equality of resources equates with equality of access to advantage. More specifically, if all resources can be transferred, then resources should be divided among individuals equally. If individuals' preferences regarding each good are different, Pareto efficiency does not exist. If the selling of resources among individuals is possible, impartial distribution as meaning of effectiveness is possible.

people at different times. Use is increased or decreased by technology, evaluation, and the work of people.

Considering the Max-min equity principle and the allocation processes, recently the community participatory approach has attracted attention as an alternative to complement market and government resource allocation (Ostrom, 1990). In a traditional way, people obtain water by self-help interaction with their community. The community shares distinctive regional information accessible only to the community, and decides management and allocation rules with consideration of their specific circumstances. If a problem arises, the community instantly brings about an efficient resolution by mobilizing local knowledge and the social capital accumulated through long-term and trusting relationships. To transfer rights to the community, a participatory approach to get mutual understanding of various stakeholders is necessary.

7.2.4 Limitations of the Participatory Approach

Self-help interactions and voluntary activities to overcome water scarcity have been implemented by individual arrangement. However, it is hard for individually arranged collective action to ensure sustainability. According to the rational choice theory and classic agency theory, collective action is motivated by self-interest, and people will join collective action only when they can expect rewards for their participation, e.g., not only pecuniary benefits but also non-pecuniary benefits such as reputation, honor, prestige, authority, power, and position. That is, collective action depends on outcome sharing. The failure of collective action has been modeled as the prisoner's dilemma and the free-rider (or moral hazard) problem (Olson, 1965; Hardin, 1968 and 1978). The prisoner's dilemma could be demonstrated by a situation with communication failure and a tit-for-tat strategy—e.g., 'I will help you if you help me'—where there a partner's treachery occurs. The point of the free-rider[5] problem could be demonstrated by a situation where a cost and benefit sharing arrangement failure occurs because of a partner's moral hazard and lack of obligation. The failure of collective action might be caused by a failure of coordination.

As a solution to the failure of collective action, Hardin (1968 and 1978) argued that the commons could be privatized or kept as public property to which rights to entry and use could be allocated. He argued that if we do not act in one of these two ways—private enterprise (controlled by market) or socialism (controlled by regulation)—we acquiesce in the destruction of the commons. However, as mentioned in the previous section, under the condition of resource scarcity, the two systems of government and market are not sufficient to overcome the exclusion problem. Ostrom *et al.* (1988 and 1990) reviewed collective action from the perspective of social capital and the participatory approach and argued that local level collective action for common-pool resource (CPR) management will be long enduring and successful under conditions of well-defined boundaries, congruence between appropriation and provision rules, effective monitoring, graduated sanctions, efficient conflict-resolution mechanisms and minimal recognition of rights to organize. As basic requirements for the success of collective actions, we can specify three factors: social capital[6] (e.g., trust and

[5]Some individuals in a population either consume more than their fair share of a common resource, or pay less than their fair share of the cost of a common resource.

[6]Social capital stands for the bonds of trust and obligation between human beings which allow non-coercive cooperation (Betts, 1998).

Table 7.1 Appropriate institutions for CPR management.

		Social Capital	
		Low	High
State Capability	Low	Private Sector	Community-based management
	High	Public Sector	Co-management with Community and State

(*Source*: Birner and Wittmer 2000).

norm), communicative action (e.g., consensus on plan or rule), and right (e.g., responsibility). Birner and Wittmer (2000) categorized appropriate institutions[7] for CPR management with consideration of social capital and state capability (see Table 7.1).

If both state capability and social capital are low, private sector management with state regulation is superior to community-based management in terms of the transaction cost[8] of collective action. Participatory management is especially suited to cases where community involvement in decision-making processes enhances compliance with resource use regulation and where equity issues need to be taken into account. Co-management may be an optimal choice where the governance structure places fewer demands on social capital and state capability (Birner and Wittmer, 2000).

Considering the exclusion problem under resource scarcity, community-based management may be an optimal choice according to the framework of Birner and Wittmer (2000). However, this type of management demands a higher level of social capital. Proper social capital can facilitate mutual understanding, mutual philia and reciprocal relationships (i.e. friendship), and mutual aid in day-to-day collaboration. Lohmann (1992) said that nonprofit, voluntary, or philanthropic behavior based on mutuality and fairness is a partial solution to the exclusion problem. The success of community-based management depends on social capital, and discussion about how to develop a reliable method to measure social capital will be important in the future.

7.3 WATER RIGHTS

There are diverse perspectives on water rights. Water right regimes can be categorized into four: customary right, riparian right, doctrine of prior appropriation, and the Dublin principle (See Table 7.2).

7.3.1 Customary Right

Water right is regulated by local customary law. Traditionally, communities have their own regulation of access to water. For example, Roman water law denied the possibility

[7]Institutions are complexes of norms and behaviors that persist over time by serving collectively valued purposes (Uphoff, 1986). Institutions are ordered relationships among people which define their rights, exposure to the rights of others, privileges and responsibilities (Schmid, 1972, p.893). Institutions determine who has the right to claim the benefits from the resource use and who must bear costs (Bromley, 1982).

[8]Transaction costs are those associated with all the exchanges that take place within an economy (North, 1990, p.27). Dahlman (1979) divided transaction costs into three types: search and information costs, bargaining and decision costs, and policing and enforcement costs. Bromley (1991) divided transaction costs into three types: information costs, contracting costs, and enforcement costs.

Table 7.2 Water right regimes.

Regimes	Concepts
Customary right	• Water is a gift from god • No individual ownership • State or community right to control but not to own water • Water allocation depends on traditional local societal hierarchies
Riparian Right	• Land ownership dependent • Reasonable use • Non-transferable right
Doctrine of Prior Appropriation	• First-come, first-served basis • Application right • Beneficial use • Transferable right
Dublin principle	• Tradable and transferable right • Participatory and democratic manner

of private ownership of running water. The Institutes of Justinian published in 533–34 held that running water was a part of the "negative community" of things that could not be owned, along with air, the seas and wildlife. Islamic law also takes this approach, which plays an important role in shaping legal rules about the use of water (FAO, 2006). The Roman law distinguished public and private water. Public water was considered to be navigable or floatable and belonged to the public or national domain. Under the nationalism of water rights, region B has no right to coerce region A. Human rights have long been subtracted from exclusive domestic jurisdiction (Teson, 1996, p.42). Their use required a government permit or authorization. That is, the Roman law denied the possibility of private ownership of running water, and the right to running water conferred no right to control its course or use, either above or below one's own land, provided its natural course and condition upon one's own land remained unchanged. Private water, which was located below, along or upon privately owned land, could be freely utilized subject to certain limitations of a statutory nature, such as servitudes and rights of way (FAO, 2006). On the other hand, Islamic law stipulates that water is a "Gift from God" and should be used for the highest level of welfare of the people. Therefore, water is owned communally by all citizens; no individual ownership can be claimed over water, and it cannot be leased out or transferred, partly or wholly. Customary rights have their roots in ancient traditions and are vulnerable to the impacts of development activities (Bruns *et al.* 2003). In more arid regions, traditional structures determine who owns water rights, and traditional arrangements for allocating the right of access to wells are related to local societal hierarchies (Arriens *et al.* 1996). The formalization process of customary waters right follows three steps: (i) context, (ii) declaration, and (iii) secure existing and longstanding use. Local ideas and practices concerning rights to water should be suitably recognized and incorporated in reforms, particularly through the empowerment of local institutions.

7.3.2 Riparian Right

In the mid-19th century, a riparian right was established, which is strongly related to landownership. Landownership was established in British society and adopted by many countries, including India and the USA. Landowners have the right to make "ordinary" use of river water for their livestock and farm assets without any restrictions, rules, or regulations. That is, the riparian right is a form of real property, a part of land law. However, the riparian right leads to diverse conflicts, such as the resistance of non-riparian residents to the riparian right, and excessive use by some riparian landowners.

The riparian right has two rules: the natural flow rule and the rule of reasonable use. The natural flow rule provides that each riparian landowner is entitled to have water flow onto his/her lands undiminished in quantity and quality. If there is sufficient flow in the stream to supply it, the upstream rights are supplied in full while the downstream rights are shut off completely. However, as the quantity of water decreases during the dry season, the burden of shortage falls on both upstream and downstream landowners. In addition, the downstream junior has the right and is entitled to insist that the upstream senior take no more than his appropriation allows. An upstream junior may divert water for his own use as long as he releases the quantity needed by the upstream senior and he may make substantial changes in the regime of the stream so long as he observes the senior's fundamental right to the use of the water (Trelease, 1963). However, senior upstream landowners cannot completely control the water flow, for example in cases of flood or drought. On the other hand, the rule of reasonable use means that the use to which the water is to be applied must be reasonable in light of all surrounding circumstances, such as the method and nature of use, the quantity desired, the quantity available, and other demands for water. This encompassed the "reasonable use" of water for domestic purposes and for the watering of livestock and, where those uses of water were made, abstraction could be undertaken without regard to the effect which they might have had on downstream proprietors (FAO, 2006). If the property is used for any other purpose, the State should have the privilege of entering and determining the riparian proprietors' estate (Alfred *et al.* 1890). There is a limit on "extraordinary water use." Some have questioned this as undesirable, and asked whether the equal sharing of the riparian right might not be better because it is even more imperfect from transferability and salability perspectives. That is, a riparian right is not very well suited to the transfer of water rights separate from the land. A riparian right cannot literally be sold as such to a non-riparian.

7.3.3 Doctrine of Prior Appropriation

The doctrine of prior appropriation handles water rights on a first-come, first-served basis, which developed in the West among pioneers who came to the vast open spaces in search of gold, land and mines. This doctrine appeared from the claims of equal importance (Trelease, 1963). The key significance of the prior appropriation doctrine is that a water right is acquired on the basis of "beneficial use," rather than on land ownership. The individual right to use water is based on the application of a quantity of water to a beneficial use.

A number of criticisms are made against the prior appropriation doctrine. One criticism is that it tends to discourage water saving by senior appropriators who know that their entitlements are relatively more secure. The definiteness of the water right in terms of "priority" and of "quantity" enables the appropriator to in effect build a fence around his water right, much as he fences his land to keep off trespassers. The feature of the prior appropriation doctrine is that it allows the senior appropriator to throw all the burden of shortage upon the junior appropriator (Molle, 2004). Therefore, the doctrine of prior appropriation has long been criticized because it promotes economic inefficiency (Tarlock, 1999). Critics seek to improve the operation of the system, by questioning how the unfortunate junior appropriators can reduce the physical insecurity of their rights. Land tenure rights, as well as trades in water rights, have long been accepted, or even encouraged, resulting in the emergence of transferable water rights. Under the doctrine of prior appropriation, the government can regulate and sanction the use of river water. For example, strict judicial or administrative enforcement of water rights can be supplemented by adaptive management schemes. Such schemes seek to meet the legitimate public and private demands of all stakeholders by a mix of creative physical solutions and water marketing strategies that equitably distribute the risks of inadequate water years among broad classes of users.

7.3.4 Dublin Principle

The Dublin principle was concluded in the run-up to the United Nations Conference on Environment and Development (UNCED), which was held at Rio de Janeiro in 1992, and has been influential in guiding the shape of water sector reforms, including reforms to water rights. In an attempt to concisely state the main issues and thrust of water management (Solanes *et al.* 1999; FAO, 2006) the UNCED provides the following.

- Freshwater is a finite and vulnerable resource, essential to sustain life, development and the environment.
- Water development and management should be based on a participatory approach, involving users, planners and policy-makers at all levels.
- Women play a central part in the provision, management, and safeguarding of water.
- Water has an economic value in all its competing uses, and should be recognized as an economic good.

The concept of the Dublin principle is that water development and management should be based on a participatory approach, involving users, planners and policy-makers at all levels. The Dublin statement suggests that water resource development and management should be planned in an integrated manner to include the following:

- Take into consideration long- and short-term water needs
- Incorporate environmental, economic, and social considerations based on the principle of sustainability
- Meet all social, economic, and legal requirements of all users
- Consider water as an integral part of the socioeconomic development planning process

Based on the Dublin principle, water right is characterized by the terms "tradability and transferability" and "participatory and democratic manner." The former characteristic is well described in the Australian Water Act. Australians were interested in tradable and transferable water rights (i.e., TWR), which began to surface in the mid-1970s, and the state of Victoria first introduced TWR in the early 1980s. In 1995 the Council of Australian Governments introduced water reform as "a system of water allocations or entitlements backed by the separation of water property rights from land title and clear specification of entitlements in terms of ownership, volume, reliability, transferability and, if applicable, quality." That is, modern water rights are not intrinsically tied to specific land plots and in an increasing number of jurisdictions they are transferable and thus may be traded on a temporary or permanent basis. A contractual water right creates an entitlement to receive a delivery of water through artificial structures. The latter characteristic is well described in terms of "the suspension or cancellation of water rights as administrative penalties." That is, water rights depend on the active management of watercourses and are mostly created on the basis of legal instruments issued by the water administration. The duration and security of water rights are important issues in administrative use or usufructuary right. Such instruments are variously described in legislation as license, permission, authorization, consent, and concession. Management of water rights requires continued activities of water stakeholders (FAO, 2006).

Under the Dublin principle, the following 3 factors are very important: (i) water administration, (ii) human resources, and (iii) financial cost. For water administration, FAO (2006) suggested five basic tasks that include planning; the organization of stakeholder fora (as described in the previous subsection); monitoring of water quality and water quantity; the issue and administration of rights, including the maintenance of registers; and the enforcement of water law and water rights regimes, including prosecution activity. Hooper (2005) suggested 9 principles of administration that include the engagement of and owners4 by relevant decision makers; improved management design; application of diverse institutional arrangements; clear definitions of the role of the river basin organization; storing advocacy; the need to prioritize action; accountability; partnerships with local government; and coordinated river basin management. As for human resources, many water projects with local residents were proposed but largely restricted by the limited human resources available. A demand-responsive approach (hereafter referred to as DRA) was proposed by Sara (1999) and followed (Deverill *et al.* 2001). The key point of DRA is the willingness to pay for the choices of community members in terms of service level, funding, and maintaining options. Households and communities must be aware of the implications of any choice they make, in terms of costs, expected participation in planning and implementation, and responsibility for operation and maintenance (Deverill *et al.* 2001).

7.4 RIGHT REALLOCATION THROUGH THE PARTICIPATORY APPROACH

The purpose of the participatory approach is to ensure coordination and consistency among stakeholders, to enhance the empowerment of vulnerable people, to stimulate local initiative, and to foster trust. The idea of empowerment and local initiative is based on morality and the strong belief that a coordinated joint voice can bring about successful collective action in consideration of vulnerable people. The coordination in the participatory approach leads to effective and equitable cost-benefit sharing and the timely delivery of service. Coordinated action improves vulnerable people's accessibility.

Three important characteristics of the participatory approach in coordination can be pointed out as follows:

- Cost Reduction
- Communicative Action
- Legitimacy and Natural Obligation

The participatory approach reduces four costs related to collective action: (i) the cost of searching for targets who should be empowered or have entitlement reallocated; (ii) the cost of reducing uncertainty about participants' capabilities; (iii) the cost of monitoring participants' responsibilities and obligations; and (iv) the cost of finding the optimal cost-benefit sharing allocation. Cost reduction is related to an increase in service production, and prevents the free-rider problem.

Social coordination by communication improves individual willingness to participate in social activities by recognizing neighbors' willingness to participate together, as well as reduces social seclusion and deprivation (or social exclusion). Habermas (1984) underlined deliberative ethic for development, arguing that communicative rationality enables people to find a new form of rationality toward agreement or consensus on plans and rules through communicative action. He said people"seek to reach an understanding about the action situation and their plans of action in order to coordinate their actions by way of agreement. The central concept of interpretation refers in the first instance to negotiating definitions of the situation which admit of consensus. ..." (Habermas, 1984, p. 86). He suggests that resolving the practical problems of vulnerable people in a degrading natural environment involves creating conditions through which they can engage in discursive deliberation free from coercion and deception. The optimistic belief is that high quality deliberation will address structural constraints on deprived people's participation (Cameron and Ojha, 2007). Habermas's idea is influenced by Kantian categorical imperative, as follows (Cameron, 1999):

- Moving to cover not only individual but also group interactions
- Linking rights to obligations
- Dealing with action at a distance through institutions
- Taking account of unequal economic, cultural and political resources/wealth
- Recognizing variations between societies and groups in thick values
- Including time as an explicit factor, taking account of duration, non-reversibility, uncertainty, and sustainability

In the case of the Kirindi Oya irrigation scheme in Sri Lanka (Molle, 2004), farmers established new allocation rules through an interactive process among them. Villagers wanted to undertake improvement to their water resources and establish their own water management systems to reallocate water in order to solve the problems of water shortages and conflicts. As in the case of the Kirindi Oya irrigation scheme, the participatory approach is associated with right reallocation. Schlager and Ostrom (1992) categorized five types of rights and the related reallocation effects.

- Access right grants entry into a defined area and specifies who is eligible to exploit a specific resource. Access right is conferred by birth (citizenship), social relations (family member), geography (local resident) or contract (license). Loss of access right may disrupt livelihood strategies, weaken social relationships among communities, and diminish sense of place.
- Withdrawal right is related to when and how individuals with access right engage in consumptive forms of resource use.
- Management right is to regulate withdrawal and control transformation and improvement.

- Exclusion right is to restrict access right. Failing to recognize "local indigenous rights" and assigning property rights to individuals or groups of users facilitate exclusive rights.
- Alienation right is to transfer management and exclusion rights to other actors.

To create the common preconditions of water right allocation—Sen's responsible choice, Sato's CPRs management by residents, Rawls's impartiality of judgment—proper public communication is needed. Sen (1999) argued that a systematic institutional framework for supporting public deliberation and its deduction or for formatting the necessary conditions affecting the result of public deliberation, is required to conduct resource management based on the proactive participation of residents. In addition, the poor are stripped of entitlement and their voice on regional resources. To address this problem and to understand the interests of diverse stakeholders regarding a regional resource (water resource), there is a need for an institutional framework on resource management based on impartiality, using the regional resource management skills and knowledge local residents have acquired up to now.

Right reallocation through the participatory approach needs a supportive legal/institutional framework for consultative roles of vulnerable users to improve their entitlement, build capability, and integrate local knowledge in managing water resources. Once a plan or a decision has been agreed upon, implementation of the water resources interventions may require external support, as communities cannot afford the investment required, due to the low productivity of their existing resources (Tod, 2003). Local participation with little funding and political authority burdens vulnerable people with a daunting political and economic challenge. In addition, the exclusion problem of vulnerable people is still not sufficiently recognized because of the social unawareness of basic rights, the plurality of administrative/legal processes, and an instinctive and understandable reluctance to engage in water resource management.

An NGO campaign for water rights, promoting recognition and engaging in monitoring activities such as mapping water resource ownership by socio-economic grouping or community could be a starting point of right reallocation through the participatory approach, which would involve discussing resource allocation and benefit distribution (Tod, 2003). Without careful monitoring of natural flows and the level of abstractions by rights holders, the security offered by a water rights scheme is lost. Such monitoring activities can be expensive, and this in part explains why in a number of developing countries formal water regimes are simply un-implemented. Endogenous attributes of the specific situation, interaction, new demand, oversimplification, conflict over existing water rights and new water rights, arbitrary decision-making in charging for water and financing of water management must be carefully considered.

Still many issues remain regarding the right to water for vulnerable users:

- Basic level of water right: The basic level of water right ensuring adequate, safe and affordable water supply should be discussed. The discussion should be based on a non-exclusion norm that one's water use for health, economic, and social benefits must never reduce any other's basic use.
- Duration of water right: When a new demand appears and an old demand disappears, the right should be created, transferred, and cancelled (withdrawal), not sustained for the property.

- Legitimacy of water right: Water rights should be provided based on the agreement of society members. There are diverse water demands, such as for residential use, irrigation, fishing, commerce, industry, manufacturing, recreation, aquaculture, navigation and water transportation, ecosystem and environment protection, hydropower, fire fighting, and flood management. Water rights for these uses must be legitimated based on evidence that the water would be used for socially desirable purposes.

7.5 CONCLUSION

This study discussed the participatory approach to water management, with consideration of the exclusion problem of poor people whose right to access water is restricted by water scarcity. Under conditions of water scarcity, there is serious concern about poor and marginal people who are easily excluded from resource allocation due to their weak entitlement. The question is how to secure such water users' access. There are voluntary challenges in providing water service to vulnerable users (Bird *et al.* 2009). However, there is no guarantee of the sustainability of voluntary water provision. With consideration of the exclusion problem, this study discussed the significance of the participatory approach in improving right to water for vulnerable users.

For this purpose, this study pointed out the problem of exclusion by water scarcity and reviewed previous water rights. With consideration of the limitations of market and government systems in water allocation, this study highlighted the importance of water reallocation through the participatory approach. The key points are water users' awareness of their real demand as willingness to pay and daily communication with neighbors to get mutual understanding. For vulnerable users, an institutional framework and effective governance for supporting and sustaining their communication and managing their resources are necessary. For support, we need to know about the structure of community governance. Finally, we need to share the methodology of enhancing the social capital which supports regional resource management by local people, and to reconsider the best practice of resource management for each country. Empirical investigations are necessary.

7.6 REFERENCES

Arriens W. L., Bird J., Berkoff J. and Mosley P. (ed.) (1996). Towards effective water policy in the Asian and Pacific region, Volume one: Overview of issues and recommendations, *Proceedings of the regional consultation workshop*, May 10–14, Asian Development Bank, Manila, Philippines.

Bird J., Arriens W. L. and Custodio D. V. (2009). *Water Rights and Water Allocation Issues and Challenges for Asia*. Asian Development Bank, The Philippines.

Birner R. and Wittmer H. (2000). Co-management of natural resources: a transaction cost economic approach to determine the efficient boundary of the state. *International Symposium of New Institutional Economics*, Tubingen, Germany.

Betts. (1998). The evolution of social capital on the origins of the social contract. *Social Contract Journal*, **8**(3), 252–261.

Bruns B.R., Ringler C. and Meinzen-Dick R.S. (2005). Water rights reform: lessons for institutional design. *International Food Policy Research Institute,* US.

Bromley D. (1989). Property relations and economic development: the other land reform. *World Development*, **17**(6), 867–877.

Cameron J. and Ojha H. (2007). A deliberative ethic for development: A Nepalese journey from Bourdieu through Kant to Dewey and Habermas. *International Journal of Social Economics*, **34**(1/2), 88–102.

Cowen M.P. and Shenton R.W. (1998). Agrarian doctrines of development: Part I. *The Journal of Peasant Studies*, **25**(2), 49–76.
Dahlman C.J. (1979). The problem of externality. *The Journal of Law and Economics*, **22**, 41–162.
Dasgupta P. (1993). *An Inquiry into Well-Being and Destitution*, Clarendon Press, Oxford.
Deverill P., Bibby S., Wedgwood A. and Smout I. (2001). Designing water and sanitation projects to meet demand in rural and peri-urban areas. *Water, Engineering and Development Centre*, UK.
Durning B. (1989). Poverty and environment: reversing the downward spiral. *Worldwatch Institute Paper* No.92, Worldwatch Institute, Washington, D.C.
Dworkin R. (1981). What is equality? part 2: equality of resources. *Philosophy and Public Affairs*, **10**(4), 283–345.
ESCAP (2009). Statistical Yearbook for Asia and the Pacific 2009.
FAO (2006). *Modern Water Rights: Theory and Practice*, FAO Legislative Study 92, Food and Agriculture Organization of the United Nations, Rome.
Gasper D.R. and Truong T.D. (2008). Development ethics through the lenses of caring, gender, and human security. *Institute of Social Studies*, The Hague.
Habermas J. (1984). *The Theory of Communicative Action,* **Vol. 1**, T. McCarthy translation, Polity Press, Cambridge.
Hardin G. (1968). The tragedy of the commons. *Science,* **162**, 1243–1248.
Hardin G. and Baden J. (1977). *Managing the Commons*, Freemen, San Francisco.
Hardin, G. (1978). Political Requirements for Preserving our Common Heritage. In: Wildlife and America, H.P. Brokaw (ed.), Council on Environmental Quality, Washington, D.C, pp. 310–317.
Hirschman A.O. (1970). *Exit, Voice and Loyalty: Response to Decline in Firms, Organizations, and States*, Harvard University Press, Cambridge.
Hooper B. (2005). *Integrated River Basin Governance: Learning from International Experience*, IWA Publishing, London.
Lohmann R.A. (1992). The commons: a multidisciplinary approach to non-profit organization, voluntary action, and philanthropy. *Nonprofit and Voluntary Sector Quarterly*, **21**, 309–324.
Molle F. (2004). Defining water rights: by prescription or negotiation? *Water Policy*, **6**, 207–227.
Nakatsuka J. (2004). World continental modeling considering water resources using system dynamics. No. 47B, Annuals of Disaster Prevention Research Institute, Kyoto University, Kyoto.
North D. C. (1990). *Institutions, Institutional Change and Economic Performance*, Cambridge University Press, Cambridge.
Ostrom E. (1990). *Governing the Commons: Evolution of Institutions for Collective Action*, Cambridge University Press, New York.
Rawls J. (1993). *Political Liberalism*, Columbia, New York.
Roemer J. E. (1986). Equality of resources implies equality of welfare, *The Quarterly Journal of Economics*, **101**(4), 751–784, The MIT Press.
Sara J. (1999). The demand responsive approach in rural water and sanitation, posted on 17 May 1999 to e-conference on DRA, http://www.mailbase.ac.uk/lists/dra.
Sato J. (2002). *The Politics of Scarce Resources: Conservation and Development in Rural Thailand*, University of Tokyo Press, Tokyo.
Sen A. K. (1981). *Poverty and Famines: An Essay on Entitlement and Deprivation*, Clarendon Press, Oxford.
Sen A. K. (1999). *Capability Approach and Poverty Reduction*, Oxford University Press, Oxford.
Siono Y. (1984). *The Structure of Value: Utility and Entitlement,* Toyo 55 Keizai Inc.(in Japanese), Tokyo.
Solanes M., Gonzalez-Villarreal F. e. a. (1999). Integrated water resources management, part I: what is IWRM? *Technical Advisory Committee (TAC)*, Global water Partnership, Stockholm, Sweden.
Schlager E. and Ostrom E. (1992). Property rights regimes and natural resources: a conceptual analysis, *Land Economics*, **68**, 249–262.
Schmid A. A. (1972). Analytical institutional economics; challenging problems in the economics of resources for a new environment. *American Journal of Agricultural Economics*, **54**(5), 893–901.
Tarlock A. D. (1999). The creation of new risk sharing water entitlement regimes: the case of the Truckee-Carson settlement. *Ecology Law Q*., **25**(4), 674–691.
Teson F.R. (1995). Collective humanitarian intervention. *Michigan Journal of International Law,* **17**, 323–370

Tod I., Parey A. and Yadav R. P. S. (2003). How can we design water resources interventions to benefit poorer households? *The Alternative Water Forum*, University of Bradford, UK.

Trelease F. J. (1963). Law, water and people: the role of water law in conserving and developing natural resources in the West. *The Wyo LJ*, **18**, 3–16.

UNICEF (2007). *Global Environmental Outlook GEO4: Environment for Development*, United Nations Environment Programme, Malta.

Uphoff N. (1986). *Local Institutional Development: An Analytical Sourcebook with Cases, West Hartford*, Kumarian Press, CT.

Zabihollah R. (1990). Water resource management. *Atlantic Economic Journal: International Atlantic Economic Society*, **27**(3), 343–352.

8 Water governance in Indonesia: from indigenous to post-colonial technology

Raymond Valiant RURITAN

Jasa Tirta I Public Corporation, Malang 65115, East Java, Indonesia. E-mail: raymond_valiant@yahoo.com and raymond@jasatirta1.co.id

Abstract

Cultural theory has been used as an approach to analyze the relationship between water and society in Indonesian history. The time frame of the Indonesian experience is roughly divided into four stages: the pre-colonial period, the western colonial period, the independent state of Indonesia, and the reformed stage of Indonesia. Analysis of the response of Indonesian water governance with regard to selected issues yields much important information. Analysis has shown that water governance in Indonesia does not provide sufficient argument in favor of one particular cultural type. One reason is that uncertainties about the various interactions between man and the environment, as well as global impacts still at large, leave room for different interpretations as to how water should be governed. Further on it can be seen that population growth and social-economic development make water governance in Indonesia a vector of political and cultural choices. Water governance in Indonesia is a culture-laden issue, reflecting political choices made in every period.

Keywords: water governance, management, Indonesia

8.1 INTRODUCTION

The main goal of this paper is to show how water is governed in Indonesia. As an archipelago strategically situated between Asia's mainland and the Australian continent, Indonesia is an important country with abundant resources. Comprising about 13,000 islands, the nation marks the equator line like a necklace; it is completely surrounded by water. Bestowed with the benefits of the monsoon spells and fertile soil, it is a paradise on water.

Rich in resources and lush in vegetation, the five important islands of Indonesia are, from east to west, Papua, Celebes, Borneo, Java and Sumatera. The population of Indonesia was estimated to be 230 million in 2010, with approximately 60% occupying Java. The main sectors of production of this country are fossil fuels, mining, manufactured goods, trading, energy and agricultural products. The Indonesian gross domestic product (GDP) for 2009 was calculated as 2,176,975.5 billion Rupiahs (based on the year 2000 prices).

Although Indonesia is blessed with water and other natural resources, adversities also loom. Besides climate and geographical disturbances like typhoons or earthquakes, human-generated problems are also emerging, especially in the water sector. For example, as water demand in certain areas rises with rapid population growth, users compete for water, depleting available resources. Even though water in Indonesia is considered abundant, elsewhere in the archipelago there has always been a scarcity. For these reasons, it is necessary to review water governance in this country. Water is a nation's asset but water governance makes it equity for the welfare of the people.

The objective of this paper is to assess and analyze water governance in Indonesia, using the cultural theory approach. Cultural theory is used because from a historical perspective, water governance is an issue inseparable from the cultural building of Indonesia. Therefore, Indonesian history will be addressed in order to classify the process by means of a distinctive cultural theory. It must be noted that during the colonial past, Indonesia was named the Dutch Indies, governed by a Governor General reporting directly to the Minister of Colonial Affairs in the Netherlands.

In order to achieve this objective, the paper is written in the following order: an introduction; problem definition; current literature information on water governance issues; an abbreviated history of water governance in Indonesia; an analysis of past and current practices; and finally the conclusion. Cultural theory is expected to be useful in order to dissect experience from the past. Voluntary actions that have contributed to secure responsible management will also be highlighted. All lessons from this paper will be useful in understanding further how water has been governed and how it will be governed in the near future.

8.2 REVIEW OF WATER GOVERNANCE ISSUES

8.2.1 Definition of Water Governance

The Global Water Partnership (GWP) defines *water governance* as the range of political, social, economic and administrative systems in place to develop and manage water resources, and the delivery of water services, at different levels of society (GWP, 2013). *Governance* in its general sense refers to the processes and systems through which a society operates. It relates to the broad social system of governing, which includes, but is not restricted to, the narrower perspective of government as the main decision-making political entity. Citing Hoekstra (2006) it can be said that water governance refers to both formal and informal structures, procedures and processes to manage water.

The main drivers behind water governance are constraints and opportunities. Four important drivers are grouped according to UNESCO (2009): demographic, social and economic drivers; technological innovations; policies, law and finance; and global climate change. Thus, water governance is akin to the capability of a social system to mobilize energy in a coherent manner for the sustainable development of water resources. The notion includes the ability to design socially acceptable public policies and mobilize social resources to support these policies, with the goal being achieved through multi-stakeholder participation.

8.2.2 Water Demand

The term *water demand* is actually not as clear a concept as it might seem at first sight. Scholars, experts and engineers use the term in different ways, reflecting three main ways to conceptualize it. The first common attitude towards water demand is to regard it as a kind of need that should be met. This approach takes population growth, economic development and the increasing demand for food as given processes that imply certain water requirements. This school of thought starts from the premise that human activities entail a certain water need, which will or will not be met. Projections of future water demand are made on the basis of growth scenarios for population, agriculture and industry, as well as assumptions with respect to efficiency improvements. This approach can often be found among engineers.

The second view of water demand is that water use is a necessity only if it is related to the fulfillment of basic needs, such as for drinking, cleaning, and food preparation (Gleick, 1996). A distinction is made between basic human and environmental needs for water and the much larger set of "need for water" to provide additional goods and services. The need for water exists independently of economic or political status, and, in principle, cannot be manipulated. Water demands above the minimum requirements are considered a luxury and largely subject to social and political desires. According to this view, economic motives and political priorities are an important factor in establishing water demand per sector. At the community level, water demand is thought to be largely a function of customs and human behavior, which may change through the improvement of environmental awareness or through, for example, the imposition of water taxes.

A third view of water demand is the economic view, in which water demand is considered subject to the price charged. Economists often speak about "effective demand" (Merret, 1997). Effective water demand is subject to the willingness and ability of people to pay for water. According to this view, water demand and supply achieve (or should achieve) equilibrium through the price mechanism. Increasing water scarcity leads to higher prices, resulting in lower demand and incentives to develop more efficient technology.

8.2.3 Water Availability

There are different approaches to defining water availability as well. One approach is to take the total annual runoff in a river basin as a measure of water availability in that basin, on the basis that fresh water is a renewable resource and the renewal rate is therefore a measure of water availability. Many authors divide the total annual runoff in an area by the number of people in that area, to obtain a measure of the available water resources per capita, either at the global or regional level (Shiklomanov, 1997). An advantage of this total runoff approach is that water availability is defined in an unambiguous way, leaving no room for dissent other than over the runoff data.

Based on the total runoff divided by the number of people, Indonesia is estimated to have an average surface flow of 15,500 m^3 per-capita annually; this is approximately 25 times the world average (Sarwoko and Anshori, 2003). However, surface water in this archipelago is not evenly distributed, especially when less populated islands such as Papua and Celebes are compared to the more populated islands of Java, Madura and Bali.

A criticism of this approach is that because it does not account for losses due to flood-runoff, runoff in remote areas, or pollution, it results in a profound overestimate. Another criticism is that because this approach considers only the possible supply of fresh water, ignoring the possibility of desalinating seawater, the approach yields too conservative a measure of water availability.

Some researchers regard the total runoff in a river basin as the upper limit to water availability and propose reductions for losses due to flooding and runoff in uninhabited areas. This approach results in a much lower water availability rate than if one were to consider the total runoff. In this approach, Hoekstra (1999), citing Ambroggi's research (1980), arrives at a figure of 23% of the total runoff and Postel's *et al.* research (1996) arrives at a figure of

31%. A further reduction could be made to account for dilution requirements and to guarantee a certain minimum runoff for maintaining river livelihoods and environmental requirements.

Another important factor of water availability is climatic variability. It would be better to use a dry year for calculations rather than an average one, to ensure that the measure of water availability also applies in dry years. The "reduced runoff approach" may have the advantage of carrying carefully balanced information on water availability, but the definitions used may give rise to many different interpretations and calculations.

In the views discussed above, water is perceived as a renewable resource. Water can, however, also be regarded as a non-renewable resource, especially in cases where humans have to rely on groundwater. A common example is when the water in rivers is too heavily polluted for human use, so that people are entirely dependent on groundwater. This leads to falling groundwater tables, most pronounced if withdrawals exceed natural replenishment, and to depletion of deeper aquifers. Additional pressures on groundwater availability are saltwater intrusion in coastal areas and other types of groundwater contamination. Two key words in this approach are therefore pollution and depletion, both processes that reduce the stock of clean, fresh water. Available freshwater stocks are now more appropriate indicators of water availability than water flows.

8.2.4 Water Scarcity

Water scarcity is a tremendous problem in many countries. Recent reports show that only 2.5 percent of the earth's water is freshwater, and about two-thirds of that is locked up in glaciers and snow. The supplies of freshwater are further diminished by at least 2 million tons of waste (human excreta, agricultural wastes like pesticides, and chemical and industrial wastes) dumped into lakes and rivers each and every day. Salinity as well as contamination from arsenic, fluoride, and other chemical toxins threatens the drinking water supply in some areas. Thus in the coming years, the combined impact of the increases in per-capita water use that commonly accompanies development, population growth overall, increased concentration of the population in urban areas, and the effects of climate change will make water more scarce in much of the world. The balance of supply to population is already uneven; Asia, with 60 percent of the world's population, has only 36 percent of the world's freshwater resources. The disparities will continue to grow. As the UNDP reported in 2004, about 20 countries face water scarcity. By 2050, water scarcity will affect at least 2 billion people in 48 countries; in the worst-case scenario, water scarcity will affect 7 billion people in 60 countries (UNDP).

However, *water scarcity* is a term that is often loosely used in different contexts. The term is used in arid countries where the dry climate is responsible for a continued state of low water availability. The term is also used to refer to temporary water shortages that occur as a result of a period of drought. If properly used, the term *water scarcity* always refers to a situation in which people have to compete over water. This often happens in arid countries (although not necessarily), but it happens in humid countries as well. Water scarcity thus relates to both climate and human activity.

One point of view is to regard water scarcity as a problem of water shortage. Water demand is seen as a given need and water supply as something that has to meet demand. If water supply does not meet demand, there is water shortage. Water demand is not only a policy issue, but it is a fact that arises from population growth and agricultural and economic developments. The actual issue is understood to be the provision of enough water of sufficient quality for the relevant sectors of society, leaving enough to fulfill ecological requirements.

The second point of view is to see water scarcity as a supply problem. In this view, water policy should aim at proper management of the physical water system, an approach found all over the world. Attention is given principally to the analysis of available water quantities and qualities and the construction of a proper water supply infrastructure. If relevant, studies should include possible effects of erosion, consumptive water use and climate change. Water pollution is described in terms of the violation of water quality standards.

The third point of view is that water availability is limited and that demand cannot continue to increase. Water scarcity is thus a demand problem. Augmentative demand is seen as the actual driving force behind growing water scarcity, with the underlying forces of population growth and economic development (Falkenmark, 1997). In nearly all parts of the world, the water utilization level is increasing, which is a signal for action in regions that have reached critical levels. Water quality deterioration is a further consequence of the increasing pressure on the water system and this problem must be solved at its roots. Wastewater treatment is not enough; wastewater production should be reduced. Solutions for water scarcity should somehow manage demand and thus human behavior.

The fourth point of view regarding water scarcity is the economic one, where it is argued that the only meaningful measure of scarcity in peacetime is the cost of the asset in question. This idea has also caught on in politics, because one of the "guiding principles for action" embodied in the so-called Dublin Statement, states that "water has an economic value in all its competing uses and should be recognized as an economic good" (ICWE, 1992). According to this view, the cost of water is the appropriate indicator to address water scarcity. If the price mechanism functions well, factors such as droughts, pollution and increasing demand will automatically account for water costs. Regretfully, solutions to water scarcity are primarily sought through privatizing water supply companies, introducing water markets and tradable water rights, and charging true costs to water users.

8.2.5 Water Management

Water management means balancing the benefits and constraints of water to achieve a desirable result. Managing water could lean toward an effort to increase the efficiency of water use (demand management) or an effort to increase the supply of water (supply management). The only exception might be for primary needs such as drinking. In this view, minimum water requirements (small but important) should be fully met, while remaining demands (large and of secondary importance) should be reduced. A reduction in water use could be achieved by, for example, increasing "water awareness" among the population and charging the user the (full) costs of water management. The price of water for primary needs should also reflect the ability of people to pay.

The technology approach, which includes such activities as building reservoirs and injecting groundwater, has become the pre-eminent engineering solution to water difficulties. The benefit of artificial surface reservoirs, for example, is the ability to stabilize variable water inflow. This can increase the stable runoff per year, which is often useful for a further expansion of water supply. Other purposes of reservoirs are for hydroelectric power generation, downstream flood control, and improved navigation. A beneficial side effect might be the recreational value of the reservoir. Nevertheless, most plans for new reservoirs today are heavily criticized, with opponents claiming that the benefits of reservoirs far from outweigh the disadvantages, such as loss of valuable land and ecosystems, (forced) displacement of people, and post-completion problems with evaporation loss, sedimentation and water quality (Scudder, 1997).

The differing points of view of reservoir advocates and opponents do not simply reflect differing results of a simple weighing of pros and cons, but can be traced to more fundamental differences in the perception of water management itself. If water management is perceived as managing supply problems, it can easily be understood why reservoirs are considered an important solution to water scarcity. Negative aspects may be serious but must be overcome. If water management is perceived as managing demand problems, however, reservoirs are not the fundamental solution, and it is therefore wise to reject their construction if there are found to be negative side effects (Pearce, 1992).

Common to bureaucrats and engineers is the economic point of view. The construction of various infrastructures for water management is not regarded as principally good or bad, and plans are evaluated mostly on the basis of a benefit-cost analysis, which might give different results in each case.

8.3 CULTURAL THEORY

The cultural theory has its roots in the anthropological research of Mary Douglas (1970) (Douglas, 2011). Further development of this theory was undertaken by Thompson, Ellis and Wildawsky (1990). Douglas introduced a group-grid typology of cultures, based on a comparison of social structures and corresponding ideas about ritual, sin and self. She argues that the character of social relations can be described along two axes: group and grid. The group axis represents the degree to which an individual is incorporated into confined units. A positive score on the group axis means that an individual strongly feels that he or she belongs to a group. The grid axis denotes the extent to which an individual's life is circumscribed by externally imposed prescriptions, or in other words, the extent to which external rules determine someone's behavior. A positive score on the grid axis indicates a high role definition, strong regulation of interactions between people and little room for individual choice. On the basis of the two dimensions, Douglas proposes to distinguish four types of social relations (Figure 8.1).

The combination of high group and high grid refers to social groups where individuals are involved with other people, but separated from them by numerous limits and boundaries. In the case of high group but low grid, all status is insignificant apart from one kind, but all status involved is in belonging to a designed group. Low group plus low grid means that individuals are free from social constraints: group organization barely exists and relaxed rules

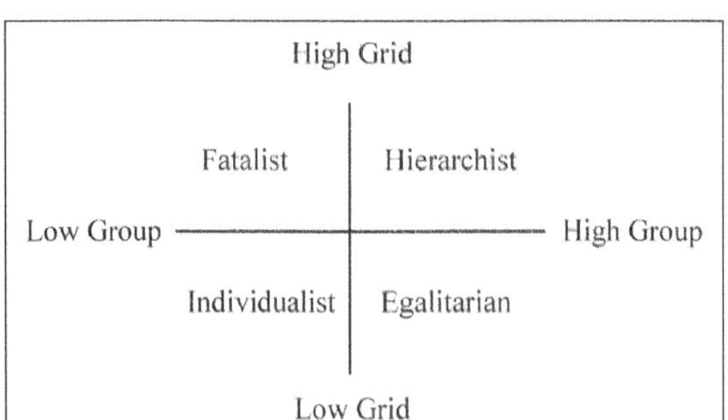

Figure 8.1 Group-grid typology of cultures (after Douglas's grid from 1974, Thompson, 1990 and Thompson *et al.* 1990).

for behavior are evident. In the last combination, low group and high grid, individuals do not belong to a circumscribed group, but they are nevertheless constrained in their relations with other people.

The group-grid typology reappears and has been further developed in cultural theory, where the four types of social structure are called "ways of life." The four ways of life are described as hierarchist, egalitarian, individualist and fatalist. To these, a fifth way of life has been added: the hermit, autonomous or intellectual way of life, where the individual withdraws from coercive or manipulative social involvement altogether. The hermit escapes social control by refusing to be controlled or to control others, but only the first four ways of life will be considered here.

Based on analysis of the group-grid typology of cultures, Table 8.1 describes the responses of the four ways of life in various terms related to social interactions.

Table 8.1 Four types of social structure, or "ways of life," according to Douglas (1974), as cited in Thompson *et al.* (1990).

Terms	Hierarchist	Egalitarian	Individualist	Fatalist
Social structure	High group, high grid	High group, low grid	Low group, low grid	Low group, high grid
Myth of nature	Perverse, tolerant	Ephemeral	Benign	Capricious
Rationality	Procedural	Critical	Substantive	Fatalistic
Knowledge	Almost complete, organized	Imperfect, holistic	Sufficient, timely	Irrelevant
Needs	Given, unmanageable	Social, manageable	Individual, manageable	Unmanageable
Resources	Scarce, manageable	Depleting, unmanageable	Abundant, manageable	Lottery, unmanageable
Management style	Control, regulatory	Preventive	Adaptive, laissez-faire	Passive
Learning style	Anticipation	Trial without error	Trial and error	Luck
Desired system's properties	Controllability	Sustainability	Exploitability	Capability
Ideal scale	Large	Small	Appropriate	No preference
Economic growth	Desirable, with conditions	Undesirable	Desirable, without conditions	Desirable, with good fortune
Desired technology	High-technology	Small-scale technology	Cheap technology	No preference
Salient risks	Loss of control	Catastrophic developments	Threats to the free market	Surprises
Risk-handling style	Institutionalization	Reduction	Taking opportunities	Acceptance

Using the group-grid typology of cultures, Hoekstra (1999) conducted further research to dissect and describe how the four ways of life types respond to water governance issues. The following description of how each way of life perceives water-related issues refers to the aforementioned study (Hoekstra, 1999). Research results are briefly summarized in Table 8.2.

Table 8.2 Responses of the four ways of life to water governance issues according to Hoekstra (1999).

Issues	Hierarchist	Egalitarian	Individualist	Fatalist
Water demand	A given need	A manageable desire	Price driven	An unmanageable desire
Water conserving technology	Large-scale technology push	Small-scale technology push	Price driven	No policy
Water pricing policy	Incremental price increase	Water taxing	Market pricing	No policy
Water availability	Stable runoff	Stable run off in inhabited areas	Total runoff or no limits	Irrelevant to individuals
Water scarcity	Supply problem	Demand problem	Market problem	Problems of individuals
Groundwater use	Inevitable	Below the sustainable level	Desirable if cost-effective	Profitable to the few
Artificial groundwater recharge	Solution to water scarcity	Should not be necessary	Desirable if cost-effective	No policy
Artificial surface reservoirs	Solution to water scarcity	Undesirable	Desirable if cost-effective	No policy
Water trade	Controlled trade	No water trade	Free trade	Trade is for the rich
Food security policy	Food self-reliance	Food self-sufficiency	Free trade	No policy
Hydrological cycle	Robust within limitations	Vulnerable to perturbations	Robust	Unpredictable
Sensitivity of the sea level	Moderately sensitive	Highly sensitive	Insensitive	Unknown
Public water supply	Incremental improvements	Basic supply for everyone	Driven by economic growth	Given to the rich
Water quality evaluation	Functional quality standards	Pristine quality as reference	Economic value	No reference
Wastewater policies	Treatment to meet standards	Treatment, decrease production	Polluter-pays principle	No policy
Flooding risks	Divergent risk levels	Equal risk principle	Economic trade-off	Risk acceptance

8.3.1 The Hierarchist Perspective

A typical characteristic of hierarchists is to regard scarcity as a supply problem. Their management strategy is to examine how they can manage their resources. Water scarcity is translated into a problem of how to increase supply in order to meet demand. Water demand is regarded as a given need, emanating from facts such as population size, economic development and need for irrigation. Water resources are available within certain limits. Stable runoff may be regarded as an appropriate measure of water availability and can be enlarged through the construction of surface reservoirs and artificial groundwater recharge. However, the ultimate limit to water availability is total runoff.

Hierarchists do not reject the further building of large reservoirs to increase stable runoff, although they recognize that the negative aspects of reservoir construction – often intangible and difficult to compare to the benefits – should be mitigated as far as possible. Groundwater use is regarded as inevitable, but the danger of over-exploitation of aquifers is recognized. Artificial groundwater recharge might be a good solution, having the additional advantage of natural purification.

Achieving food security is an important aim, but hierarchists do not consider it necessary to attain this goal by striving for food self-sufficiency. Countries should aim at some sort of food self-reliance, where food needs can be met through a combination of their own production and a stable trading environment.

Hierarchists are willing to strive for more efficient water use, but one should not have great expectations of programs aimed at the development or introduction of water-conserving technology, particularly not if one expects changes to come from collective efforts to introduce small-scale water-conserving technology. People should rather aim to develop and introduce high technology on a large scale (on the supply side, not the consumer side).

Hierarchists are not inclined to push water prices strongly in the direction of real costs (or market pricing), because a rapid increase in water prices would disturb socio-economic stability to an unacceptable extent. As their ultimate goal, hierarchists aim for a situation where water charges fully cover operational and maintenance costs. It is not considered fair that water consumers should have to repay all investment costs. An argument for this position is the importance of investments in the public water supply for improving public health. In the case of irrigation, one might say that irrigation investments stimulate the general economy and that hidden taxes are often already imposed on farmers through price controls for agricultural products.

For hierarchists, public water supply and proper sanitation facilities for everyone are desirable goals.

Hierarchists, who perceive nature as tolerant, consider that disturbances such as global warming, land use changes and consumptive water use will alter the hydrological cycle to some extent, but not uncontrollably. It is assumed that disturbances can be assimilated as long as they do not reach critical levels. An issue such as wastewater treatment becomes important if water quality standards are not reached. Standards can differ by type of water source and type of intended use. Risks of flooding are, if possible, regulated by the formulation of maximum acceptable risk levels and the improvement of dykes or other defenses to conform to these levels. Acceptable risk levels vary for different areas, from relatively high in undeveloped areas to comparatively low in highly developed areas.

8.3.2 The Egalitarian Perspective

Egalitarians are prudent in assessing water resources, and take account of temporal and spatial variability. Stable runoff in inhabited areas may be an appropriate measure of water availability, but other indicators of water use are also essential, such as the actual decline of groundwater tables and the remaining amount of high-quality water.

Water scarcity is regarded as a problem caused by growing water demand and pollution. The solution lies in the management of human needs. Water demand is seen as a manageable desire that can be changed by policy incentives and shifts in social customs and preferences. Applying small-scale water conservation and reuse technology can lower water-use intensities in all sectors. The egalitarian is more sensitive to communal programs to introduce new technology than to an increase in the water price.

Nevertheless, to accommodate the environmental consequences of excessive water use, egalitarians advocate that such impacts be included as a tax in the price of water. However, everyone should have access to water to fulfill basic needs, which means that water should be free to people who otherwise would not have it.

Egalitarians attach great importance to equity; access to safe drinking water and sanitation facilities for everyone is a principal policy goal. It would be typically egalitarian to promote a second International Drinking Water Supply and Sanitation Decade. The (first) International Drinking Water Supply and Sanitation Decade (1981–1990) was also an egalitarian initiative.

In the egalitarian view, too much bureaucracy has meant that goals have been far from achieved. From a hierarchist point of view, the failure to reach goals has instead been the consequence of several kinds of inevitable social and political constraints, such as insufficient involvement by women and political resistance to cost sharing.

According to the egalitarian, the fragile dynamic equilibrium of the water balance is easily disturbed by human activities. Intensive water use, human-induced temperature change and deforestation may considerably affect stable runoff and the sea level. Fertilizers and household and industrial waste will not only remain in some hot spots, but will spread throughout the world. According to egalitarians, wastewater should, as a matter of principle, be treated before disposal. Bodies of water should if possible be returned to their pristine quality, i.e. the quality before significant human disturbance.

Egalitarians are strongly opposed to the further building of large reservoirs, arguing that the social and ecological costs of dam construction far outweigh any possible benefits. Groundwater use should be reduced to stop the overexploitation of aquifers. As a measure of an acceptable level of groundwater withdrawal, one should not only look at natural recharge (which might result in an overestimate), but also at the actual effect of withdrawals on water tables. Artificial groundwater recharge is not regarded as a real solution, because the water would have to be taken from surface waters that are vulnerable to overexploitation as well.

Water trade in any form is considered undesirable, because water is seen as public property. Importing or exporting water-intensive products (trade in virtual water) is undesirable as well. Transfer of water between different river basins is rejected from an ecological point of view. In principle, countries should strive for water and food self-sufficiency. In the egalitarian view, the risks of flooding should be reduced first in areas where risks are highest (equal risk principle).

Egalitarians are most concerned with the protection of less developed regions, where poor but densely populated areas are exposed to regular flooding. In the case of increased flooding frequency as a result of land cover changes, erosion or climate change, preventive strategies are preferable to defensive strategies.

8.3.3 The Individualist Perspective

The perspective of individualists largely coincides with the need to emphasize the economic point of view: water is an economic good and should be managed as such. For individualists, all options to improve water supply conditions should be realistic, provided that they are cost-effective. Efficiency improvements that reduce demand are often profitable, as they save not only water but also money.

However, extending the resource base – for instance through exploitation of untouched aquifers or increasing desalination capacity – can be profitable as well. The right mixture of water supply and demand management will be the beneficial function.

Individualists consider total runoff the proper measure of water availability. If water recycling and desalination techniques become more efficient and economically feasible on a large scale, water might even become an unlimited resource, so that the problem will no longer be one of availability but one of the efficient exploitation of water. The hydrological variability of water in time and space is not a real limiting factor to individualists, who regard free trade as the ultimate solution to carry water and water-intensive products to the demand areas.

Water demand is determined by the price mechanism: higher prices as a result of increased scarcity will lower demand and stimulate the development of more efficient technology. If new water-conserving techniques become cost-effective, they will replace older techniques, prices will decrease and demand will rise again. Individualists discourage subsidies on water (which is at present a common practice all over the world) (Young, 2005).

Water prices should be established by market mechanisms. In cases of high water scarcity, high-tech options for water supply (e.g. desalination) could be stimulated by government institutions, but always on a payback basis.

Individualists do not pursue an active policy in public water supply and sanitation, because they believe that economic development will increase public water supply and sanitation coverage adequately. Economic development is even regarded as a prerequisite for water supply and sanitation improvements. According to the individualist point of view, wastewater treatment is an economic trade-off. Application of the "polluters pay" principle will force polluters to treat wastewater if this is preferable to paying for the damage caused by pollution (which has to be expressed in financial terms in some way). The value of a body of water of a certain quality depends on its economic value.

In the individualist view, reducing or accepting risks of flooding is an economic trade-off. In line with their perception of nature as robust, individualists tend to regard possible disturbances of the hydrological cycle as of minor importance. If intensive water use, land use changes or global warming have some effect on the hydrological cycle, the resulting changes will occur slowly enough for people to adapt.

8.3.4 The Fatalist Perspective

According to fatalists, life is unchangeable. Because of so many uncertainties, changes to the hydrological cycle can in practice be regarded as unpredictable.

Whether people are provided with enough clean water or not is seen rather as a matter of individual luck than as a matter of national or regional water shortage or abundance. Questions about whether water is a finite or infinite resource or how water availability should be measured are considered academic and irrelevant to individual people.

Water scarcity is seen as a problem of individuals. According to fatalists, water is always given to the rich, either in water-poor or water-rich parts of the world. The poor seem always to lack adequate water supply and sanitation under all circumstances. This lack creates a downward spiral that hampers the poor's opportunity to improve their welfare.

Water demand is regarded as an unmanageable desire, whether it is satisfied or not. Water trade is possible only for rich people in power and does not benefit the poor. Risks of flooding are accepted as fate, and therefore have to be endured because fatalists do not feel they can reduce them.

Fatalists are not in favor of any particular management strategy, which means that their management strategy essentially comes down to doing nothing, merely coping with whatever situation evolves. In the fatalist view, people are unable to control the future, and even if they could, interests diverge and strategies would counteract each other to such an extent that the net result would be a lottery. An increase in water prices or the introduction of water taxes would not make sense, because one could question who would profit and who would suffer.

The crisis in water and sanitation is – for the fatalist above all – a crisis for the poor. Almost two in three people lacking access to clean water survive on less than $2 a day, with one in three living on less than $1 a day. More than 660 million people without sanitation live on less than $2 a day and more than 385 million on less than $1 a day (UNDP, 2007).

Several studies show that it is the urban poor who will spend more of their money and a higher proportion of their income on water due to the average tariff on clean water service (SEAWUN, 2007). Chances are that increasing prices would make life worse for the poor, without affecting the rich who can pay easily. According to fatalists, water inequality is a living risk that need not be battled.

Improving water conservation awareness among people through education is not regarded as a very useful policy either. According to the fatalist point of view, the people who might learn most from these kinds of programs are often not those members of society who are in a position to conserve much water. To the contrary, people in a position to conserve water are those who are already well educated but probably unwilling to give up their privileged use of water (such as for swimming pools, Jacuzzis and sprinklers in the garden).

8.4 INDONESIAN HISTORY OF WATER GOVERNANCE

The history of water governance in Indonesia can be traced back from the early Indian-influenced kingdoms in the sixth century on the island of Java, to the Indo-china and Islamic and western colonial periods, and finally into the independence phase of the Indonesian state.

8.4.1 Pre-Colonial Hydraulic Civilization

Water is an abundant resource for the Indonesian archipelago, where monsoons reign during half the year, bestowing rain on the volcanic-rich islands. However, Indonesia has inherited both the benefits and adversities of water. Most surface water in the archipelago is provided by the monsoon rain; however, the monsoon season is also associated with flash floods, inundations and landslides. Albeit during the dry season, in areas where less water can be captured and stored in the watershed, drought is often found. As an example, drought analysis shows that in recent years (1991, 1994, 1997 and 2002) the island of Java has endured several low rainfall periods. Food production is much affected: the decrease in rice-grain production was about 784,000 tons (about 3%) in 1991; 1,751,000 tons (6%) in 1994 and 558,000 tons (2%) in 1997. The decrease in 1997 was the worst, amplified by the Southeast Asian financial crisis, which prevented the government from being able to import rice from the world grain trade (Pangesti and Budinetro, 2003).

Concentrating on the more populated islands in the archipelago like Sumatera and Java, we can see that the interaction between water and society has resulted in a "hydraulic civilization." It this type of civilization, water governance (especially in terms of irrigation for wet-farming activities) is related to a centralized governance system – where the ruling class plays a distinctive role. Even this term from Karl A. Wittfogel (1896–1988) may not fully explain the dynamic relation between water and society, but it can depict the fact that water management requires societies to use collective forces within themselves. The "hydraulic hypothesis" argues that the development of Egypt, Mesopotamia, India, China and Pre-Columbian societies was blocked because of the need to irrigate vast surfaces for agriculture. Water control and distribution spawned authoritarian centralized empires and sprawling bureaucracies, both deeply hostile to change.

The Indonesian experience is rather different. As in most Southeast Asian countries, water engineering feats in Indonesia keep pace with water governance. However, water governance in Indonesian history is mostly related to specific issues, like providing water for irrigation or flood control. Specifically, irrigation was the starting point of water governance.

Evidence of early agricultural activities in Indonesia was found at East Java. These agricultural activities did not require irrigation. Archeological evidence at various caves and rock-shelters in the karst area suggests widely spread pre-historic agricultural activities known as the "Sampung Culture," dating to between 6 to 3 millenniums before the Common Era (BCE). Sampung is a limestone cave near Ponorogo, East Java, where the archeologist L.J. van Es in 1926 conducted investigations on pre-historic agricultural activities. Later research by Stein-Callenfels (1932) in East Java confirmed his findings. As such pre-historic agricultural activities can be found as far away as Situbondo, Bojonegoro and Tuban, they were grouped and simplified as the "Sampung Culture" (Tanudirjo, 1997). Cultivation was mainly of bulbs, such as taro (Colocasia esculenta), yam (Dioscoreaceae), arrowroot (Maranta arundinacea) and others.

Evidence of wet-farming and irrigation has been found since the first Indian-influenced kingdoms in the island of Java. Irrigation is related to the introduction of rice cultivation (Oryza sp) in the first century BCE by the Indo-Chinese migrating populace (Setten, 1979). Rice cultivation by wet-farming is thought to have originated in Assam, India, and spread along the southeast Asian peninsula, along with Hinduism (Arief, 2003). However water governance practices in the past can be traced – often faintly – only by artifacts or written information left by the rulers, not by the commons.

In the year 417 CE in the northern part of West Java, King Purnawarman from the Taruma – the oldest recorded kingdom in Indonesia – inaugurated the completion of the Chandra-bhaga River rehabilitation. Completion of this river rehabilitation was anointed in the Tugu Inscription found near Jakarta, the capital of Indonesia. Tugu is one of 7 inscriptions left by the

Taruma Kingdom that ruled the lower lands of West Java for almost 3 centuries (approximately 358–690 CE). The river is presently known as Bekasi River (bhaga evolved linguistically to Bekasi). The Chandra-bhaga River rehabilitation is aimed at protecting farmland from inundations and providing accessibility to surface water for irrigation purposes. The river rehabilitation – about 11 km long – cannot fit Wittfogel's "hydraulic hypothesis" or the implied idea of "oriental despotism," as there is lack of information on the role of the ruling class in rehabilitating the river. However, Taruma was a rich agricultural based kingdom; it occupied the fertile lands in the lower part of Cisadane, Cianten, Ciaruteun and Cikasungka Rivers. Most likely, the Chandra-bhaga River was rehabilitated by the surrounding communities under the guidance of local potentates and religious leaders (Figure 8.2).

Figure 8.2 Remnants of an old canal at Plaosan Temple, Central Java, from the 9th CE, used to divert water from a nearby river probably for sacred use (courtesy of Arief, 2003).

After the decline of the Taruma Kingdom, there was a brief period with no artifacts to provide information on water governance in Indonesia. However, it is strongly argued that the Indian influence – including the cosmic view of the world in the Hinduism pantheon – made water an important aspect of religiosity. This was confirmed by the Tukmas Inscription found at Grabah, Magelang District in Central Java, which bore praise to the holiness of a spring and based on the paleographic analysis of the inscription, and is moderately dated 650 CE (Sarkar, 1969). The Tukmas Inscription is also referred to as one of the earliest Hindu cosmic influences in Southeast Asia. The full inscription reads *(itant) uśucyamburuhānujātā/kwacicchi lāwālukanirgateyam/kwacitprakīrnnā śubhaśī tatoyā/samprasratā m(edhya) kariwa gańgā* , translated as (from) this shining lotus flower/flows the spring of purity/gushing through sand and stones/the sacred cool water like the Ganges (river) (Soehadi and Soekarto, 1986).

An important term to note here is that in this inscription water is associated with religiosity. This archaic theme could be traced in various inscriptions left by later rulers. Giving water this association emphasizes the way rulers exercise their power on natural resources by placing them under the "sacred canopy." Even if we understand that voluntary actions – especially of peasants and other water-using members of the society – are imporant in securing the water supply, the ruler's power remains important.

Water sources – springs as well as rivers – were considered sacred at that time. The Wanagiri Inscriptions I and II (dated 904 CE) from Balitung, the ruler of the older Mataram

Kingdom, depicts the Bengawan Solo River as *sang hyang mahawan,*, or the auspicious lord. As Soekmono (1995) quoted from Stutterheim's research (1934) regarding this inscription, Balitung was said to *kamalir muang kamulān muang parahu*, or to build a simple house and a boat as a ferry across the Bengawan Solo River at Wonogiri District. This inscription provides us information about the navigational purposes of the river at that time.

Another inscription dated 934 CE by Sindok, another ruler of the Mataram Kingdom, recalls the river as *luah prasiddha*, or sacred flow. Other evidence of how water was treated as sacred in the cosmic view of the early Indonesian kingdoms can be seen in the spatial design of the kingdom's court. Most courts were put in a diagonally correct position between the mountain and the sea, with the river crossing the imaginary line between them. The connection of these important nature elements assures the cosmic reliance and strong belief in the royals at court. The remaining royal courts in Cirebon, Surakarta and Yogyakarta are silent proof of this fact.

Water engineering feats reflect the technology that was used. In the Indonesian experience, especially in Java where artifacts are found, the applied technology is quite advanced.

Weir or *prisé d'eau* in the old Javanese language is called *dawuhan*, which mainly consists of a construction crossing the river in order to raise the water level and thus enable water to be abstracted for irrigation purposes. The Harinjing Inscription (921 CE) in East Java described an event where a local potentate named Bhagawanta Bhari led his people to construct a *dawuhan* at the Serinjing River for irrigation purposes (Lombard, 2000a). Upon completion of the *dawuhan*, Sindok, the ruler of the Mataram Kingdom, blessed the populace by releasing them from taxes, thereby ordaining the land a free-estate (*sima*). Other inscriptions like the Bakalan (934 CE), Kamalagyan (1037 CE), and Jiwu (1486 CE) also cite the construction of *dawuhan* at various rivers in East Java for irrigation purposes.

Here, it is important to note the existence of the Kamalagyan Inscription, dating from a later period when East Java was ruled by King Airlangga. The benefice charter of Kamalagyan, issued in the delta of the Brantas River in East Java, concerns the repair of a major dam (*dawuhan*) at Kamalagyan and a related dike (*tambak*) at Waringinsapta. This inscription is unique in the surviving corpus of the Old Javanese language, in that it alone concerns a water control project that involved direct royal action. It is also the only case on record in which the primary purpose of the project was not religious, although the community involved had a connection with a local water temple named the Lake of Dharmma (*ranu ring dharmma*) (Boomgaard, 2007). The Kamalagyan Inscription, however, shows that during King Airlangga's (991–1052 CE) reign in East Java, there was a turning point in the matter of water governance. Initially, the construction of various water infrastructures was conducted by communities under the guidance of the local potentates. The rulers' role was to ordain or simply to commemoratively anoint the construction results. However, under Airlangga there was a shift, as the ruler used his royal corvée rights to rebuild the reservoir after the local populace had failed in their attempts to make repairs and then approached him for aid. This is an important turning point, proving that as early as the 11th century the "hydraulic civilization" had begun.

An interesting form of infrastructure from the classic period prior to the western colonization of Java is the underground channels, one of which was reportedly found by Setten van der Meer (1979) near the District Purworejo, Central Java. During a flood in 1861, overflow from the Kragilan Channel unearthed an older underground channel from the 13th or 14th century. This finding was not extraordinary, as other archeologists had excavated underground conduits at different places around Java. These underground channels were named *arung* in old Javanese. At the Jalatunda, Simbatan, Karuman, Palawijen and Wanajati sites – all at the hills around Mount Penanggungan (1,653 meters) in East Java – such conduits were found as remnants of the King Airlangga period (1019–1042 CE). The same findings were noted by Mac Laine-Pont at the Pikatan and Kromong River, Mojokerto District, East Java. Underground channels with an average size of 2.5 m in width and 3.0 m in height were constructed in the late Majapahit Kingdom era (middle of the 15th century) to carry water to the opposite side of the mountain, which had less surface runoff.

Navigational canals were another feat, traditionally constructed to execute the transport by water of goods and people. One famous canal system is found at Trowulan, the former capital of the Majapahit Kingdom (1293–1478). These canals can be traced by their contrast to the surrounding common soil, as seen from aerial photographs. The systematically built canals dissecting the city of Majapahit are the results of wise deliberation and advanced civilization showing concern for the environment. Remaining traces of canals can be detected at various places within the area of the old capital. The width of the canals is about 25 meters and they have a depth of approximately 4 meters.

From archaeological evidence found at the site, it can be concluded that the capital city of the Majapahit Kingdom at the Trowulan Site was built through a process of deliberation and accomplished by thorough planning, with detailed and modern architecture indicating local wisdom in caring for the environment. This provides proof of the accumulation of knowledge and ideas by a sophisticated civilization of the ancestors of Indonesians in the 12th and 14th century CE. With dimensions of approximately 9 km by 11 km, the capital was situated at the border between Jombang and Mojokerto District, East Java.

The network of canals in the Trowulan Site crossed the city almost perpendicularly. Apparently Majapahit city was developed on a chessboard pattern, shaped by the relatively straight and perpendicular canals stretching from north to south and west to east. The course of the canals was not necessarily parallel to the earth's north-south magnetic axis, but were shifted slightly to the right, clockwise in the Cartesian quadrant. It appears that the canals were adjusted to geographical conditions. Judging from the distance of the canal grids in the map, in the western part, the north-south canals were located relatively closer to each other than those built in the eastern part. This shows that in the zones where the canals were relatively close, the area was utilized for settlement, the city center and the king's palace. Meanwhile, the east-west canals, built straight and intersecting the central part of the canal system, provided evidence that there was a link for socio-cultural activities connecting the eastern, western, northern and southern parts to the central part of the city. The canals were also linked to a network of roads built parallel to the canals either on one or both sides of the canals. A distinctive name for canals in the old Javanese terminology is *weluran*, a term adopted into the Indonesian language as *saluran*.

It is clear that the canal system and water structures built in the Majapahit era served also as irrigation facilities for agriculture and were used to channel water into man-made reservoirs. Trowulan had five reservoirs, namely the Baureno, Kumitir, Domas, Temon, Kraton and the Kedungwulan. In addition to these, Trowulan had three man-made ponds closely positioned, namely the Balong Bunder, Balong Dowo, and the Segaran Pond. These reservoirs functioned as water storage for domestic purposes.

Water governance during the pre-colonial time was basically conducted by the village-level government. At large-scale canal networks such as in Trowulan, the capital of Majapahit, a more sophisticated administration was appointed by the rulers. However, specific job titles in irrigation and water services, such as *matawmak* (those responsible for weir construction and maintenance), *hulu wuatan* (bridge), *pangarung* (underground channel) and *hulair* (water distribution), were appointed by the populace (Subroto, 1997). The system can be freely described as decentralized, within the boundaries of the village or free-estate (*sima*). The regime of the ruling class can be described as sacramental, ordaining and anointing the management process, while operational aspects were handled by village leaders and local potentates.

However, it can be seen that the voluntary participation of the society is important under these development schemes. The willingness of the people to enact the water supply systems plays a role that fits into the hierachical category.

8.4.2 Western Colonial Period

The Dutch entered Indonesia for trading purposes as early as 1596, when Cornelius de Houtman arrived in Banten, West Java. But it was not until the beginning of the 20th century that the Dutch finally colonized the archipelago that was later reclaimed as an independent state of Indonesia. However, water engineering expertise from the Dutch was not shared until the late colonial period that began in the 19th century after the Dutch recaptured Indonesia as a colony after the Napoleonic War (1803–1815).

Eventually, the Dutch tried to introduce certain engineering projects in the areas surrounding their trade strongholds. Water engineering projects were connected to economic endeavors. As an example in 1647, two officers of the Vereenigde Oost-Indische Compagnie (VOC), or the Dutch Indies Company, were given a permit to dig canals for navigational purposes. VOC was a chartered company established in 1602 in Amsterdam, when the States-General of the Netherlands granted it a 21-year monopoly to carry out colonial activities in Asia. It was the first multinational company in the world, and the first company to issue stock. It was also arguably the world's first mega-corporation, possessing quasi-governmental powers, including the ability to wage war, imprison and execute convicts, negotiate treaties, coin money, and establish colonies. However, due to corruption, this company was liquidated in 1798.

Regarding these two officers, François Caron built the Ancol Canal and Johannes Ammanus built the Ammanusgracht, both connecting the market in Batavia with the surrounding farmland. For the same reason, in the year 1648, Phoa Bing Ham, the "Chinese captain" of Batavia (serving 1645–1663), who was also a wealthy merchant, dug out the Molenvliet (the mill canal) in Batavia. The Molenvliet canal was used to transport goods and connect the economic activities of the old city of Batavia (named Kota) with the residential area (Weltevreden), or what is now the Tanahabang Sub-District (Lombard, 2000b).

Furthermore, in 1682, Vincent van Mook, a Dutch merchant, obtained a permit from the Governor General of the Dutch Indies to connect Batavia (presently Jakarta) and the hinterland through an excavated channel for navigation purposes. Approximately 18 km long, the channel connected the Angke River in Batavia with the Cisedane River, and upon completion was named the Mookervaart Canal. In addition to its navigationel purpose of transporting agricultural products from the hinterland to the harbor in Batavia, this channel also served as a water source for an irrigation area of approximately 34,000 *bouw*. The main products from the hinterland and its surrounding areas were sugarcane, rice and vegetables.

The Mookervaart Canal inlet was directly connected to the Cisedane River, thus making sedimentation a problem within the canal. After contemplative study, as noted by Ravensteijn (1997), in 1919 the engineers van Breen, Polderman and Dumont proposed a plan to secure this inlet. The plan consisted of closing the old inlet of the 17th century, providing a *schutsluis* to enable ships to enter or leave the canal and backwash gate to provide water to the canal. This plan was executed in 1934 and completed in 1936, converting the Mookervaart Canal into an irrigation channel adjacent to the Tangerang Irrigation Works (Ravensteijn, 1997).

By the year 1748, due to the rapid development of private-agriculture estates in the hinterland of Batavia, the Governor General of the Dutch Indies, Baron van Imhof, granted a permit for landlords to construct an irrigation channel abstracting water from the Ciliwung River at Katulampa near Bogor District, running northward, before finally draining into the Sunter River at Batavia. This irrigation channel was named the Oosterslokkan, or the eastern channel, when it was completed in 1758. A succession of landlords, the first of whom was S. Johan van Wimmercrantz, were charged with the operation and maintenance of this channel. The Oosterslokkan was later integrated into the Ciliwung-Katulampa Irrigation Works in the 19th century (Wirosoemarto, 1997a). In conjunction with this channel, another was constructed between the Cisedane River and the Western Flood Canal near Manggarai.

Another effort by the Dutch in terms of water engineering was to settle drainage problems within Batavia, the capital of the Dutch Indies, by developing canals resembling the *gracht* system in Amsterdam or Rotterdam. Remaining *gracht* can be found in the northern part of Jakarta, such as what is now called the Sunter River and the Pasar Baru River. However, these feats were mostly sporadic and did not take into account modern planning methods employed nowadays.

A new realm was generated when in 1801 the civil engineering polytechnic at Delft was established. Graduates from this polytechnic served as engineers to deal with infrastructure problems in the colonies, including Indonesia. One of the first modern engineering approaches introduced by these engineers involved the less well known Sedadi Canal constructed under the Vice Governor General for the Indies, J.H. Daendels, 1810–1811. Daendels was appointed by Napoleon Bonaparte (1769–1821), the Emperor of France, after Napoleon had successfully occupied the Netherlands in the fierce Napoleonic War. The objective of the Sedadi Canal was to connect cities along the northern shore of Java (Semarang-Demak-Kudus-Juana) by inland water navigation. This canal mainly uses water from the Serang River. Due to limited knowledge of the surrounding geo-morphology and hydrological conditions, this *prauwvaart* canal did not work well, and was soon abandoned.

By 1830, the Governor General of the Dutch Indies introduced an enforced plantation policy that obliged peasants in Java and Sumatera Island to render one-fifth of their land for hard crops like tea (Camelia sinensis), quinine (Cinchona sp), coffee (Coffea sp) and chocolate (Theobroma cacao). This policy enriched the colonial treasury but devastated the food production that sustained the life of commons. After a series of small famines throughout Java and Sumatera, 1848–1849, a large-scale famine hit the Semarang, Grobogan, Demak and Rembang areas along the northern shore of Java. Approximately a quarter of a million people died. Ashamed of this devastating outcome, the Dutch Indies colonial government appointed a "Water and Famine Commission" (*Water en Voedingsnood Commissie*) by 1850.

The commission endorsed the colonial government to invest more in land development for food production and to improve irrigation works to secure production. The response was the first large-scale irrigation project under the colonial government, 1852–1858, the construction of Glapan Weir at Tuntang River, in the western part of Semarang, Middle Java. Two large canals were constructed, taking water from the Glapan River, to irrigate approximately 15,530 *bouw* or 10,770 hectares of rice plantation.

While the Glapan Weir was being constructed, the Dutch Indies colonial government took the remarkable step in 1854 of establishing the Water Branch within their administration. Whereas irrigation and water-related issues had theretofore been handled by engineers embedded in the local Dutch Indies administration – the *eerst-aanwezend water ambtenaar* (first engineering officer) – after the establishment of the Water Branch, these resident regency-level engineers could report simultaneously to the colonial government in a centralized way.

It is quite fascinating that even before the great famine and its aftermath, by the year 1832, an *eerst-aanwezend water ambtenaar*, the engineer C. van Thiel – stationed at Besuki Residency – constructed the first modern weir at the Sampean River, in Pasuruan, East Java. The weir was simple: constructed from rafts of teak wood (*tectona grandis*) filled with andesitic stones, it served as a water abstraction point for the related agricultural land. This weir has been reconstructed over time, and even now still serves irrigation purposes. Adjacent to the vast agricultural development in the Pasuruan area, which produced rice, corn (*zea mays*) and tobacco (*officinarum tabacum*), Sampean Weir was further developed by the colonial government, and in 1872 could serve 13,600 *bouw* of irrigated land. An additional overflow channel was added by the engineer S. Dik (1847), improvements to the permanent weir were made by J. Schumm (1871) and a flood channel was added by the engineer M. Ypelaar (1884).

Meanwhile, in the Brantas River Basin, East Java, another improvement was carried out. In order to satisfy water requirements for more than 16 sugar cane factories and plantations in the delta of this river, in 1847 a water gate was constructed at Mlirip, Mojokerto to bifurcate the Brantas River into the Surabaya River (running to the north) and the Porong Canal (running to the east). In order to abstract water for the sugar cane industry and plantations, a weir was constructed near Mangetan at the Surabaya River. However, this effort was considered insufficient, so that by 1852 another barrage had been constructed at the Brantas River at Lengkong, near Mojokerto. The main function of this barrage was to supply more water to the delta and to regulate high flows into the Porong Canal.

Initially the gates of Lengkong Barrage were designed as portable locks; however, due to operating difficulties, the barrage was reconstructed in 1857, with a simplified gate type. Later, in 1972, with Japanese technical assistance, the Lengkong Barrage was completely reconstructed and moved to a new site (Figure 8.3).

This multi-purpose engineering feat marks the Dutch water plan in the Brantas River Basin. Along with this, further water infrastructures were built in the basin, most aimed at securing irrigation for sugarcane plantations and rice estates. At around the beginning of the 20th century, the Dutch had built about 6 prominent free intakes along the Brantas River, thus securing water for approximately 40,000 hectares of fully irrigated land. In between the colonial government plans, some small-scale efforts to manage water were found also in the Brantas River Basin, such as the initiative of Aria Natadiningrat, head of the Malang District, who in 1854 constructed the Kedungkandang Weir as a *prisé d'eau*. After its completion, this weir was destroyed in a flash flood in 1867. It was neglected for quite some time, before the colonial government replaced the weir at a newly selected site in 1904, reviving the designated irrigation area. However, this plan was simple compared to the zealous efforts by the colonial government to enlarge irrigation areas within the Brantas River Basin (Wirosoemarto, 1997b).

As the requirement for a more technocratic system emerged, by 1866 the Dutch Indies colonial government had reformed the Water Branch and transformed it into a directorate under the Directorate General of Public Works. The Public Works consisted of the Water Directorate (or *Waterstaat)* and the Road and Transportation Directorate (or *Verkeer*). Thus, the Directorate General of Public Works at that time was also named Department van V en W (with V for *Verkeer* and W for *Waterstaat*).

Figure 8.3 The old Lengkong Barrage at Mojokerto, taken from upstream in 1920 (Courtesy: KITLV, Leiden).

Water governance was executed through a broader approach within the administration. However, as irrigation projects in the Dutch Indies began to emerge, dissatisfied with the current way of administration, by 1885 the colonial government had established a special Irrigation Brigade to study, design and execute irrigation projects in the Indies. The first action of this brigade was to initiate, by the year 1886, the Serayu Irrigation Project in the Banyumas and Bagelen District, Central Java. This project replicated the previous success of irrigation development in the Brantas River delta. Furthermore, by the year 1872 the *Demaksche Waterwerken*, or the Eastern Semarang Irrigation Plan in Central Java, also commenced. The plan comprised construction of Sedadi Weir to abstract water at Serang River for approximately 21,330 hectares of wet-cultivation. In accordance with the Java overall irrigation plan, the *Demaksche Waterwerken* was extended in 1897 by constructing the Wedung Weir to serve an additional 5,500 hectares of irrigation area.

With the huge expansion of irrigated areas – especially on the island of Java – by the year 1889 the Irrigation Brigade had gained more resources and evolved into the Irrigation Section directly under the Directorate General of Public Works of the Dutch Indies colonial administration. With more resources at hand, by 1890 the brigade published its first meticulous irrigation plan for the island of Java. Irrigation plots all over the island were evaluated and assessed, thus creating the first rational plan for water governance in Java. In an effort to create better water governance and solve water engineering problems, by the year 1897 an engineering quality assurance commission (*Rentabiliteitcommissie*) was established by the Irrigation Section. The purpose of this commission was to assure a better technocratic approach in the field of water governance. By the year 1900, total irrigated land by modern irrigation methods was approximately 100,000 hectares (close to 5% of total wet-farming land).

Technical irrigation expanded rapidly in the first quarter of the 20th century as result of the huge investment in infrastructure by the colonial government. Irrigated land was about 5% of the overall cultivated farmland in Java in 1900, and it approached 25% (or 678,000 hectares) in 1925 (Happé, 1939). Vast irrigated areas in Java facilitated agricultural development. Water governance under the efficient colonial administrative system contributed to the production of rice, sugarcane and maize, important products of Java at that time.

However, among the ruling class, in this case the kings or what was left of them from the royal line of Java, there were also opportunities to develop entrepreneurial activities. Using Dutch Indies "water expertise," members of this class developed various ponds to dispense water for paddy and sugarcane cultivation ventures within their lands. For example, the Mangkunegara Court built the Plumbon near Eromoko in the Wonogiri District, and the Delingan (Tirtomarto) in the Karanganyar District, all in Central Java, while the Kasunanan Court built the Cengklik in the Boyolali District, and the Mulur in the Sukarharjo District. All ponds were aimed at irrigating the sugarcane estates, owned by the royal court. Governance of these ponds was unique. As they were constructed with funds from the court and technical assistance of the Dutch, water management for irrigation was brought to concession with the royal government (a *waterstaat* office for each court).

It is important here to note as well a huge irrigation plan by the *Waterstaat* in the valley of the Bengawan Solo River. This mega-project, named Solo Valleiwerken, had been planned since 1852 by the *eerst-aanwezend water ambtenaar*, the engineer A.A.V. Lederboer, but it was not until the year 1870 that the plan was evaluated, and the Water Branch dispatched J.L. Pierson to conduct a preliminary study and design. By the year 1881 a plan was drawn up. First, the Bengawan Solo river mouth near Sembayat, Gresik District would be reconstructed so as to be equipped with a flood canal at Pelangwot, near Babat Sub-District, that could bifurcate flood water during the monsoon season and lessen inundations in the lower areas. Second, a weir would be constructed at Ngluwak, near the Ngrahu Sub-District, Bojonegoro, to abstract water for approximately 223,000 *bouw*, (or 158,000 hectares), of irrigation to a system of channels that would be a total of 900 km long. The main irrigation channel would be 30 meters in width and could be used for navigational purposes. In 1892 the construction process started, with a budget of 15 million guilders. However, due to cost overruns and disputes within the colonial administration, by end of 1897 the Director General of Public Works, de Meijer, brought the project to a halt. Most of the construction equipment was sold, the remaining buildings were left, and the approximately 120,000 hectares of acquired land was handed to the government of East Java Province. The Bengawan Solo Valleiwerken was an unsuccessful water infrastructure project in the Dutch Indies.

Besides addressing irrigation expansions, the Dutch Indies colonial government began to pay more attention to flood control. Flood by nature is a persistent problem for Indonesia, especially on Java, where flood prone areas are mostly the food bowls and agglomeration centers: Batavia, Bandung, Semarang, Surakarta, or Surabaya. The city of Surakarta, for example, was established at the bend of the famous Bengawan Solo River in Central Java. The city had endured flood, and a large flood in February 1861 inundated the city for weeks, a disaster which spurred the construction of a dike system surrounding the city, funded by the Dutch Indies colonial government at the request of the royal families ruling Surakarta (Figure 8.4).

Figure 8.4 Flood (1869) by Raden Saleh Sjarif Boestaman (1811-1880) one of the most well known painters and a pioneer of modern Indonesian art.

On February 24–25, 1902, another big flood hit the Semarang, Kudus and Jepara area on the northern shore of Java, devastating not only settlement areas, but also inundating hundreds of thousands of hectares of farmland, creating a period of famine after the flood. This flood incident brought up the first modern flood control plan, namely the Semarang Flood Plan in 1908. The plan was to control the Serang and Babalan Rivers by, among other things, establishing a "stoney-type" gate: the Wilalung Barrage. Flood control was further established as a policy of the Directorate General of Public Works, when on October 6, 1918, another large-scale inundation submerged Batavia, the capital of the Dutch Indies. The engineer Herman van Breen was assigned to complete a flood control plan for the capital. After a studious survey and design process, he came up with a plan to construct two flood canals that would embrace Batavia and cut the large discharges from the upstream area by directly dispatching them to the sea. Breen was given the budget to construct the most urgently needed one, and by 1922 completed the West Flood Canal (in Indonesian: Banjir Kanal Barat) securing most of Batavia from inundations.

Despite the fascinating and satisfying result in Java, the development of water resources infrastructure on other islands in Indonesia started much later. It was not until the beginning of the 20th century that the Dutch Indies colonial government initiated modern irrigation and flood control on other islands. Attention to irrigation in Sumatera Island was aroused only after the colonial administration realized that Sumatera –famous for its plantations of tobacco and coffee – could not sustain itself in terms of food sufficiency. By the year 1913, the Irrigation Section initiated the Bah Korah I and Bah Korah II Project in North Sumatera, as the first modern step toward irrigation in Sumatera. However, agricultural development in Sumatera could not cope with the growing food demand. Wirosoemarto (1997a: 331) notes that rice imports to West Sumatera, for example, in the years 1915, 1916, and 1917 were 91,000, 127,000 to 146,000 tons, respectively, and by the year 1929 had reached 181,000 tons. Irrigation development was also introduced to other islands much later, not until 1925 for the Jeneberang Irrigation Plan in South Celebes and the Bolango Irrigation Plan in Gorontalo, North Celebes.

While earlier water projects were mostly aimed at building individual works for one or a few purposes, this changed around 1900 with the construction of multipurpose reservoirs and other water works. These large reservoirs were built in large rivers for irrigation, electricity production and other purposes. The Aswan Reservoir, designed by Sir William Willcocks and completed in 1902, was the first big reservoir in the modern world. Willcocks dreamt of a system of reservoirs serving several purposes, but his creation was in fact mainly for the production of electricity. The Aswan High Dam followed only in 1970. The international management of the Nile waters was organized long after 1902 and is still an issue to be definitively settled.

Engineers in the Netherlands East Indies started building reservoirs shortly after Willcocks, but their first reservoirs were not really multipurpose. Paradoxically, however, they followed upon an early multipurpose project, namely the Solo Valley project on East Java. This was a large-scale water works project which comprised works for irrigation and drainage and, in addition, included provisions for shipping and the drinking water supply.

The plan also provided for a diversion of the river to the Java Sea in the north in order to prevent its gigantic quantities of sediment from flowing into the Straits of Surabaya and endangering the shipping to and from the Surabaya harbor. This multipurpose project, however, was soon stopped for financial and technical reasons. The alternative was a small reservoir plan, which provided for the improvement of small Javanese reservoirs as well as the construction of new ones. Some big reservoirs were also part of the plan and two were realized in 1917 (Prijetan, 9 million cubic meters) and 1933 (Pacal, 60 million cubic meters).

8.4.3 Post-Colonial Period

After a brief period of uncertainty caused by the Asia-Pacific War (1942–1945) and the independence struggle (1945–1949), the Dutch Indies was replaced by the independent state of Indonesia. Of course, the Indonesian government understood the importance of public works, and the importance therefore of maintaining the existence of the previous Directorate General of Public Works under the Dutch Indies colonial government. This directorate was changed briefly into a Japanese Occupation office for public works, *Kotubu Bunsitsu*, and at the first ministerial cabinet established by President Soekarno – the Indonesian independence leader and also an engineer himself – was reinstalled as the Ministry of Public Works (2011). The first appointed Minister of Public Works was Abikusno Tjokrosujono (1897–1968). As he was a political figure, his office was held rather briefly (approximately three months) before he was replaced.

Following the pattern of the Directorate General of Public Works, the Ministry consists mainly of three elements: water, road and housing (settlement) sections. These sections were transferred into Directors under the Minister, thus upholding a new level of management.

However, in the matter of water governance, there was no abrupt change. The Ministry of Public Works maintained its administrative function during the period of uncertainty. The planning and technocratic function did not reside. Reservoirs were also prepared elsewhere in Java, the most ambitious plan being the three multipurpose reservoirs in the Citarum River proposed by engineer Willem Johan van Blommestein. His "Welfare Plan" comprises the following concepts: creating large reservoirs in the western part of Java for hydro-electric power generation and huge irrigation schemes, reclaiming the Segara-anakan near the Cilacap District, and constructing a sea dike at the northern part of West Java. These large reservoirs were only built after Indonesia became an independent Republic in 1945/1949. The first one (Jatiluhur, 3 billion cubic meters) was realized in 1969, with French technical assistance.

Later, in the beginning of the 1960s, another huge development scheme was prepared in the Brantas River Valley, where, with Japanese financial assistance, a cascading series of reservoirs was constructed, namely Selorejo (completed in 1972, 54.6 million cubic meters), Karangkates (1973, 253 million cubic meters), Wlingi (1978, 5.2 million cubic meters), Lodoyo (1983, 5 million cubic meters), Bening (1984, 24.8 million cubic meters) and Wonorejo (2000, 106 million cubic meters) (JICA, 1998). These multipurpose reservoirs provide flood control at the upper stream, generate electricity, and distribute water for various purposes such as industrial bulk water, domestic bulk water and irrigation.

When they started to improve and build reservoirs in the Solo Valley, Dutch East Indian engineers were inspired by the numerous small Javanese reservoirs in the area. Apart from indigenous knowledge, these engineers used knowledge gained from reservoir building in Europe in connection with industrial activities in Belgium and Germany, among others. Consequently, a Dutch tradition of reservoir construction and management came into being in which additional knowledge from especially the British colonies and the United States of America was absorbed. Van Blommestein improved the financial-economic basis of his Citarum reservoir plan, for instance, by combining loss-making irrigation with profitable electricity production, which he learnt from the 1930s Tennessee Valley Authority project in the USA (Ravensteijn, 2007).

An important characteristic of water governance during the post-colonial period was the utilization of basin-wide planning principles in developing the affirmed water resources potential within a river basin. The first ever basin-wide plan was initiated by the Japanese Government, under the war reparation fund, which developed the Brantas Plan in 1959, rewritten and officially confirmed as a plan by the Ministry of Public Works under the Indonesian Government in1961. The Brantas River in East Java at that time threatened a large fertile land area in the lower basin with flooding, due to the enormous quantity of volcanic ash deposited in the river bed after the eruption of Mount Kelud (1,731 meters), an active volcano situated in the middle course of the river. Flood control and erosion control, therefore, were most urgently needed in this basin, as addressed by the first plan developed by a consulting firm, Nippon Koei K.K. from Japan. The proposed plan was to construct at Karangkates a reservoir large enough to control flooding during the monsoon and distribute water during the dry season (Nippon Koei K.K., 1961).

The plan comprised a more comprehensive approach to addressing ongoing problems in the basin, such as flood and drought, and related objectives for the welfare of the inhabitants. Further, the plan drew more attention to the idea of comprehensive plans. After the huge flood of March 6, 1966 in the upper Bengawan Solo River Basin, another basin-wide plan was prepared, which evolved into the first master plan for the Bengawan Solo River Basin, 1974.

The development of basin-wide plans for river basins was then adopted by the Ministry of Public Works, reflecting a new approach to the planning, designing and executing of water infrastructure. Basin development planning became more inclusive and more strongly based on a sound understanding of the rivers' boundaries.

This integration used the famous maxim "one river, one plan, one integrated management," widely adopted since the United Nations coined it in the early 1960s. Various master plans for river basins have been carried out in accordance with the concept expressed in this maxim.

Finally, to cope with the recent reform process of 1998, the Indonesian Water Law of 1974 was replaced by the new Water Law No 7 of 2004, which enables more cooperation by agencies within a basin and also coordination of policies dealing with water governance. For instance, the new law does not allow farmers to pay for the water service they receive, and also opens new possibilities for river basin organizations to develop themselves.

8.5 ANALYSIS OF PAST AND PRESENT WATER GOVERNANCE

An analysis of past and present water governance practices in Indonesia is done based on the cultural theory adaptation in water resources management, using Douglas' (1974) approach, as developed by Hoekstra (1999). First, important issues in water management are inventoried. Second, the past and present conditions of Indonesia are translated into columnar stages. Third, using information from various reports and research, water governance at every stage is evaluated and categorized in accordance with Hoekstra's assessment. Fourth, based on previous reviews, every stage's response to water governance is classified into one of the four types of culture. Finally, at the fifth stage, a conclusion is drawn.

Basic results of the analysis are drawn in Table 8.3 and Table 8.4. From the four columns, i.e. pre-colonial period, western colonial period, independent state and present reformed state, interesting conclusions can be drawn. Basically, the social structure within each period affects policies, perception and implemented actions by various parties involved in water governance.

At the pre-colonial period, Indonesia was marked by lower population density. The economy was characterized by simple agricultural efforts, mainly consisting of typical slash-burn cultivation with a limited amount of wet-farming. The ordained kings and priests played an important role in the cosmic view of the world. The commons shared their resources in an egalitarian way, if not described as fatalistic. Water was governed mainly at the village and free-estate levels under the auspices of landlords and local potentates; the ruling class played the more or less sacred role of ordaining and anointing the feats executed by the commons. Water governance in that period did not fully develop into a "hydraulic civilization" until (at least) King Airlangga, who, according to the Kamalagyan Inscription, initiated larger schemes of water infrastructures using his royal powers (Ras, 1994).

It is interesting to see how at that time water infrastructures were funded by (mostly) the commons. Organizing construction activities and providing funds for operation and maintenance were mostly borne by the users. An earmarked fund may not be easy to prove, but users were responsible for financing the construction and operation of the related structure. A certain water fee was recorded in the Jiyu Inscription (1486 CE) that stipulated every water user from the Trailokyapuri Weir was obliged to pay a tax of 3, or up to 8 chin of rice and a certain amount of gold. This fee was a precursor to the application of the cost recovery scheme for water services (Dwijanto, 1997).

Shifting into the western-colonial period, the Indonesian landscape endured unabridged changes. Population grew especially in agglomeration enclaves; trade relationships flourished; intensive agricultural efforts were made, providing more benefits to capital owners and less to the commons. The classic colonial portrait of the Dutch Indies was achieved during the period of 1602–1816.

Western colonialism took a different route after the Napoleonic War (1803–1815) ended. Europe's wealth was depleted due to the war, and replenishing their treasury became an important objective. This led into a devastating *cultuurstelsel* that "drained the natural resources" of the Dutch Indies. As the second enlightenment took place in the first half of the 19th century, colonialism took a different course along with the political change in the

Table 8.3 Analysis of various water governance issues (first part).

Issues	Pre-colonial Period	Western Colonial Period	Independent State	Present Reformed State
Water demand	A manageable desire, mainly because of the low demand [Egalitarian]	A given need, price driven issue for colonial expansion purposes [Individualist-Hierarchist]	A given need, need for development expansion [Hierarchist]	A given need, but for distinctive regions it becomes a price driven issue [Individualist-Hierarchist]
Water conserving technology	No significant policy, however possible to have small-scale push by indigenous knowledge and wisdom [Fatalist-Egalitarian]	No policy, destruction of watershed due farming practices is evident due to colonial policies [Fatalist]	Evolving policies, small-scale actions. Even though watershed destruction is becoming a growing threat [Egalitarian]	Large-scale push, grand design to conserve more by various means. Watershed destruction, global warming threats [Hierarchist]
Water pricing policy	Water taxing and royal corvée to construct public water infrastructures. [Egalitarian]	Government/private financing for water infrastructures; royal corvée whereas until 1836 forced labor is still evident [Individualist-Hierarchist]	Water taxing as the way of the government to finance water infrastructures [Egalitarian]	Water tax and water fee with incremental billing methods used [Hierarchist]
Water availability	Irrelevant to individuals, mainly because of the low population [Fatalist]	Stable runoff in inhabited areas, mainly due to the limited areas of growth [Egalitarian]	Stable runoff [Hierarchist]	Stable runoff [Hierarchist]
Water scarcity	Problems of individuals [Fatalist]	Supply problem [Hierarchist]	Supply problem [Hierarchist]	Demand and supply problem [Egalitarian-Hierarchist]
Groundwater use	Below the sustainable level, mainly because of the low population [Egalitarian]	Below the sustainable level, mainly because of the limited areas of growth [Egalitarian]	Higher pressure on groundwater, considered inevitable [Hierarchist]	Inevitable, for distinctive areas of growth it is desirable when cost is lower [Individualist-Hierarchist]

Table 8.3 Analysis of various water governance issues (first part). (*Continued*)

Issues	Pre-colonial Period	Western Colonial Period	Independent State	Present Reformed State
Artificial groundwater recharge	No policy, because groundwater use is below the sustainable level [Fatalist]	No policy, because groundwater use is limited to certain areas of growth. [Fatalist]	Solution to water scarcity [Hierarchist]	Desirable if cost-effective [Individualist]
Artificial surface reservoirs	Solution to water scarcity [Hierarchist]	Solution to water scarcity [Hierarchist]	Solution to water scarcity [Hierarchist]	Solution to water scarcity, but should be cost-effective. [Individualist-Hierarchist]

Netherlands. This change is abbreviated as a "mutual welfare concept," driven by social ethics and technology development. However, colonialism remained as a faithful servant of the new capitalism; the Dutch Indies still served as what Deventer cites as "the cork wood the Netherlands floats on." As researched by Ricklefs (1981), it can be noted that between the early 1830s and the late 1870s, the Dutch Government had secured revenues totaling almost 820 million guilders from its colonial territories in southeast Asia; by 1860 it is estimated that the Dutch Indies provided a third of the Netherlands' state revenue (Tan, 2008).

Water governance in the Dutch Indies evolved rapidly during western colonialism, especially after social ethics and technology development were introduced. In terms of water governance, rational planning and efficient administration are two important pillars left to the independent state of Indonesia as a legacy of the Dutch Indies colonial administration.

The Indonesian state has adopted these important pillars, but ameliorates water governance in different ways to suit to the needs of a diverse archipelago marked by unbalanced population density and striding economic movements. Turmoil in the first 25 years of the Indonesian

Table 8.4 Analysis of various water governance issues (second part).

Issues	Pre-colonial Period	Western Colonial Period	Independent State	Present Reformed State
Water trade	No water trade, mainly because of the low demand [Egalitarian]	No water trade, mainly because of the restricted development scheme [Egalitarian]	No water trade, mainly because of the absence of related regulation [Egalitarian]	Controlled trade, limited within the regulations [Hierarchist]
Food security policy	Food self-sufficiency, however, limited population to feed at that time [Egalitarian]	Free-trade, colonial implication on food production is controlled by the market [Individualist]	Food self-reliance, in certain sense trade of crops [Hierarchist-Individualist]	Food self-reliance, in certain sense trade of crops [Hierarchist-Individualist]

Table 8.4 Analysis of various water governance issues (second part). (*Continued*)

Issues	Pre-colonial Period	Western Colonial Period	Independent State	Present Reformed State
Hydrological cycle	Vulnerable to perturbations, water management affected by the cosmic view of the world [Egalitarian]	Robust, classic approach in planning [Individualist-Hierarchist]	Robust within limitations, classic approach in planning [Hierarchist]	Robust within limitations, detailed planning using water balance modeling [Hierarchist]
Sensitivity of the sea level	Unknown [Fatalist]	Moderately sensitive, the first references to ebb and tide levels for design purposes [Hierarchist]	Moderately sensitive, sea water level incorporated into design and planning [Hierarchist]	Moderately sensitive, sea water level incorporated into design and planning [Hierarchist]
Public water supply	Basic supply for everyone, mainly because of the low population [Egalitarian]	Basic supply for everyone, mainly because of the moderate population [Egalitarian]	Incremental improvements, due to population growth and competing demands [Hierarchist]	Incremental improvements and economy-driven services [Individualist-Hierarchist]
Water quality evaluation	No reference, mainly because no significant problems are evident [Fatalist]	No reference, even if significant problems are evident, no enforced standards [Fatalist]	Functional quality standards, evolving into regulations [Hierarchist]	Functional quality standards, desire to adapt economic instruments [Individualist-Hierarchist]
Wastewater policies	No policy, mainly because no significant problems are evident [Fatalist]	No policy, even if significant problems are evident but no public regulation enforced [Fatalist]	Treatment to meet standards [Hierarchist]	Treatment to meet standards, even if standards are becoming harder to achieve. [Hierarchist]
Flooding risks	Divergent risk levels, by constructing simple flood control measures [Hierarchist]	Divergent risk levels, by constructing more complicated flood control measures [Hierarchist]	Divergent risk levels, by constructing complicated flood control measures [Hierarchist]	Divergent risk levels, but with some improvement to divide risk [Egalitarian-Hierarchist]

Government was a part of this evolution; however, after an abrupt political change in 1966, water governance was conducted in a well-ordered manner, incorporating planning prudency and systematic development. Water engineering feats were diligently recorded in this period, mainly emphasizing a hierarchist approach towards water engineering. This condition was evident for approximately 30 years, until corruption and political instability caused another abrupt change in the country in 1998.

Both political and administrative reform as a result of this change has completely altered water governance. The structured approach as the signature of the hierarchical cultural type was questioned. Difficult questions also arose: what are water rights, who is responsible for managing water, who are stakeholders, where can the people participate and how can water be managed? With regard to the reform process, a new Water Law was enacted, in a compromise between the hierarchist system, individualist pressure (especially due to economic reform to a more liberal approach) and egalitarian demands of the commons.

Based on the analysis, it can be concluded that water governance in Indonesia has evolved in stages since the pre-colonial period to the era of the independent state of Indonesia. During the pre-colonial period, water governance was mainly attributed by the egalitarian or nearly fatalistic type of culture. As western colonialism invaded Indonesia, water governance was initially attributed to individualistic endeavors, but as the colonial administration took form, it shifted into a hierarchist type of culture. The colonial was able to translate its administrative power into creating a strong group and grid type of culture. After Indonesia sailed into its independence state, the hierarchist approach in water governance was carried over by the newly established Indonesian Government. This hierarchist type remains in water governance practices, showing that the strong grid type remains favored by the Indonesian Government.

However, the Indonesian Government changed its policies in 1998 and conducted governmental reforms. Among the products of the reform was the new Water Law of 2004, which aims to accommodate shifts in water governance. The role of the government is at present partially deconstructed, opting for a more de-centralized and de-concentrated form of governance. This implies that the government has to change the way they manage resources. The related Department of Public Works cannot act as a complete provider for any problem regarding water, but it must transform itself into an enabler.

On the other hand, the people are requesting more control of the available resources. During the reform period they also had to change themselves into a looser grid-relationship. People are not looking the same way as before: more regional interests have been built. The loosening group role and the untying of the grid type can be reflected in the change from the more hierarchic approach in the "independent state" stage towards the adaptation of a loose individualist type of culture. In a certain respect, water governance in Indonesia is also moving towards a more egalitarian way.

In the egalitarian way, the power of the collectives, when they speak with a common voice, shall have more power to pressure management into action, such as increasing services or reducing fees, or even changing allocation.

As water resources management crosses sectors and areas of interest, the egalitarian way will require cohesiveness of action to maintain sustainability of function and the benefit of water resources. As a platform for stakeholder liaison, cohesiveness of action and an advisory mechanism for water-related basin-level development and management, a coordinative and consultative mechanism is required to integrate the interests of various sectors, territories and stakeholders in water resources.

According to the Indonesian Water Resources Law No.7 of 2004, coordination of water resources management must be established at the national level (National Water Resources Council), at the provincial level (Water Resources Council or other name) and at the basin level (Water Resources Management Coordination Team).

Prior to the establishment of the Coordination Team, in order to avoid conflict among water users in the various basins in Indonesia, especially in the densely populated Island of Java, provincial water resources committees have been established. These committees, responsible to the Governor, are supposed to be coordination bodies where decisions on all management aspects of water resources (planning, implementing, supervising, controlling and funding) in their respective areas are made. From this perspective, it can be said that the hierachial approach is still evident.

As the more individualist-hierarchiest approach gains momentum, under the Water Law No 7 of 2004, the coordination bodies established throughout the basins in Indonesia are expected to provide more space for users and stakeholders to voice their interests.

The role of government is changing from that of a developer to that of an enabler, one which provides a place for corporate-type river basin organizations to run water services and engage in operation-maintenance activities in certain basins, under a system based on water service fees. These river basin organizations will be playing a bigger role in the future, rendering services separated from regulatory aspects, but still curbing the individualist drive that looms whenever a "free-market" approach is used in managing water.

8.6 CONCLUSION

Based on the previous analysis, it has been demonstrated that past and present practices in water governance can be reviewed using the existence of cultural types. These cultural types will differ in their underlying basic values, beliefs and assumptions, making them a useful tool in understanding the circumstances that affect policies.

Cultural theory was used as an approach to analyze the relationship between water and society in Indonesian history. In order to provide a synchronous time frame, the Indonesian experience is roughly dissected into four stages: the pre-colonial period, the western colonial period, the independent state of Indonesia, and the reformed stage of Indonesia. Analysis of the response of Indonesian water governance with regard to selected issues proves much important information.

It can be stressed that water governance in Indonesia does not provide sufficient argument in favor of one particular cultural type. One reason is that uncertainties about the various interactions between man and the environment, as well as global impacts still at large, leave room for different interpretations of how water should be governed.

However, it can be seen that population growth and social-economic development are two aspects that make water governance in Indonesia a vector of political and cultural choices. Water governance in Indonesia is a culture-laden issue, and reflects political choices in every period.

8.7 ACKNOWLEDGMENTS

The author acknowledges the Board of Directors of Jasa Tirta I Public Corporation for their permit to do research for this paper. Much gratitude is also conveyed to Bertha, Benehad and Benrachum for their patience and understanding during the author's long hours of labor.

8.8 REFERENCES

Arief S.S. (2003). Merenungkan Kembali Pengelolaan Sumberdaya Air Berkelanjutan Berbasis Budaya Lokal. *Proceedings of the Seminar held by Food and Agriculture Organization (FAO) and Indonesian National Planning Board*, Jakarta, Indonesia, vi–xvi.

Boomgaard P. (2007). *A World of Water: Rain, Rivers and Seas in Southeast Asian Histories*. Koninglijke Instituut voor Taal, Land en Volkunde (KITVL) Press, Netherlands, 248–249. (ISBN 978-9067182942).

Douglas M. (2011). A history of Grid and Group Cultural Theory. University of College, London, England. Retrieved from http://projects.chass.utoronto.ca/semiotics/cyber/douglas1.pdf (Retrieved on January 2, 2011).

Dwijanto D. (1997). Perpajakan pada Masa Majapahit dalam Sartono Kartodirjo (ed.), *700 Tahun Majapahit (1293–1993)*. East Java Provincial Office for Tourism, Surabaya, Indonesia, 217–234.

Falkenmark M. (1997). L'interaction de la société et du cycle hydrologique: un cadre conceptuel pour une approche globale. *Joumal-des Sciences Hydrologiques*, **42**(4), 421–454.

Gleick P. H. (1996). Basic Water Requirements For Human Activities: Meeting Basic Needs. *Water International*, **21**, 83–92.

Global Water Partnership (GWP). (2013). *Integrated Water Resources Management Tool Box B.01*. Retrieved from: http://www.gwptoolbox.org/index.php?option=com_tool&id=9 (Retrieved on May 29, 2013).

Happé P.L.E. (1939). Eenige beschouwingen over nevloeiingswerken op Java en Madoera. *De Ingenieur in Nederlandsch-Indië*, **6**(1), 13–26.

Hoekstra A. Y. (1999). Appreciation of water: four Perspectives. *Water Policy*, **1**, 605–622.

Hoekstra A. Y. (2006). *The Global Dimensions Of Water Governance: Nine Reasons For Global Arrangements In Order To Cope With Local Water Problems*. Water Value Research Report Series 20, UNESCO-IHE Institute for Water Education, Delft, Netherlands.

Indonesian National Statistics Bureau. website: http://www.bps.go.id/aboutus.php?tabel=1&id_subyek=11 (Retrieved on February 11, 2011).

International Conference on Water and the Environment (ICWE). (1992). *The Dublin Statement on Water and Sustainable Development*, Dublin, Ireland.

Japan International Cooperation Agency (JICA). (1998). *Development of the Brantas River Basin, Cooperation of Japan and Indonesia*, Tokyo, Japan, 108.

Lombard D. (2000a). *Nusa Jawa: Silang Budaya Bagian III: Warisan Kerajaan Konsentris*, PT Gramedia Pustaka Utama, Jakarta, Indonesia, 19–21.

Lombard D. (2000b). *Nusa Jawa: Silang Budaya Bagian II: Jaringan Asia*, PT Gramedia Pustaka Utama, Jakarta, Indonesia, 251.

Merret S. (1997). *Introduction to the Economics of Water Resources: An International Perspective*, UCL Press, London, England.

Ministry for Public Works (2011). http://id.wikipedia.org/wiki/Daftar_Menteri_Pekerjaan_Umum_Indonesia (Retrieved on February 16, 2011).

Nippon Koei K. K. (1961). *Comprehensive Report on the Kali Brantas Overall Project, Government of the Republic of Indonesia*. Ministry of Public Works and Power, Jakarta, Indonesia, 3–14.

Pearce F. (1992). *The Dammed: Rivers, Dams, and the Coming World Water Crisis*. The Bodley Head Press, London, England.

Pangesti D. R. and Budinetro H. S. (2003). Drought and Flood Anticipation in Java, Indonesia in *Proceedings of the 1st International Conference on Hydrology and Water Resources in the Asia Pacific Region 2*, Asia-Pacific Association of Hydrology and Water Resources, Tokyo, Jepang, 775–780.

Ras J. (1994). Geschiedschrijving en de legitimiteit van het koningschap op Java. *Bijdragen tot de Taal, Land en Volkenkunde*, **150**(3), 518–538.

Ravensteijn W. (1997). *De Zegenrijke Heeren der Wateren: Irrigatie en Staat op Java 1832–1942*. Delft University Press, Delft, Netherlands.

Ravensteijn W. (2007) *Colonial Technology and the European Unification: The Case Of Integrated River Basin Management*. Retrieved from: http://www.histech.nltensphase2/meetingsfinlandCD (Retrieved on December 8, 2010).

Rogers P. and Hall A. W. (2003). *Effective Water Governance*. TEC Background Papers 7, Global Water Partnership, Stockholm, Sweden.

Ricklefs M. C. (1981). *A History of Modern Indonesia*. Indiana University Press, United States of America.

Sarkar H. M. (1969). South-India in old Javanese and Sanskrit inscriptions. *Bijdragen tot de Taal, Land en Volkenkunde,* **125**(2), 193–206.

Sarwoko A. and Anshori I. (2003). Keterpaduan pengelolaan sumberdaya air untuk pendayagunaan yang berkelanjutan. *Proceedings of the Seminar* held by Food and Agriculture Organization (FAO) and Indonesian National Planning Board. Jakarta, Indonesia, 7–22.

Scudder T. (1997). Social impacts of large dams in Dorcey (ed.); *Large Dams: Learning From The Past Looking At The Future*, The World Conservation Union and the World Bank. Washington D.C. United States of America, 42–68. (ISBN 0-8213-4028-X).

Setten van der Meer N.V.C. van (1979). *Sawah Cultivation in Ancient Java: Aspects of Development During the Indo-Javanese period, 5th to 15th Century*, Australian National University Press, Canberra, Australia.

Shiklomanov I. A. (1997). *Comprehensive Assessment of the Freshwater Resources of the World: Assessment of Water Resources and Water Availability in the World*, World Meteorological Organization, Geneva, Switzerland.

Soehadi M. and Soekarto M.M. (1986). *Laporan Penelitian Epigrafi Jawa Tengah 37, Proyek Penelitian Purbakala Jakarta*, Departemen Pendidikan dan Kebudayaan, Jakarta, Indonesia.

Soekmono R. (1995). The Javanese Candi: Function and Meaning, Studies in *Asian Art and Archeology*, E. J. Brill, Leiden, Netherlands, 69.

South-East Asian Water Utilities Network (SEAWUN) (2007). *Data Book of South-East Asian Water Utilities 2005*, SEAWUN and Asian Development Bank.

Stein-Callenfels P. V. van (1932). Note Preliminaire Sur Les Fouilles Dans L'abricious Roche de Guwa Lawaa à Sampung. *Hommage du service archeologique des Indes Neerlandaises au 1ere Congres des Prehistoriens à Hanoi*, Albrecht, Batavia, Indonesia, 16–32.

Subroto P. (1997). Sektor Pertanian Sebagai Penyangga Kehidupan Perekonomian Majapahit dalam Kartodirdjo (ed.): *700 Tahun Majapahit*, East Java Provincial Office for Tourism, Indonesia, 155–176.

Tanudirjo D. A. (1997). Pertanian Majapahit sebagai Puncak Evolusi Budaya. In: 700 Tahun Majapahit, Kartodirdjo (ed.), East Java Provincial Office for Tourism, Indonesia, 131–152.

Tan M. G. (2008). *Etnis Tionghoa di Indonesia*, Yayasan Obor, Jakarta, Indonesia, 116.

Taylor G. (1979). *International Encyclopedia of Social Sciences 18*, Collier Publishers Ltd., London, England, 812–813.

Thompson M., Ellis R. and Wildawsky A. (1990). *Cultural Theory, Westview Press, Boulder, Colorado*, the United States of America.

United Nations Development Program (UNDP) (2007). *Water Governance for Poverty Reduction: Key Issues and the UNDP Response to Millenium Development Goals*. New York, United States of America.

United Nations Development Program (UNDP) (2007). *Human Development Report 2006, Palgrave Mac Millian*, New York, United States of America, 3–25. (ISBN 0-230-50058-7).

United Nations Educational, Scientific and Cultural Organization (UNESCO). (2009). *Water in a Changing World*, UNESCO Publishing, Paris, France. (ISBN: 978-9-23104-095-5).

Wirosoemarto S. (1997a). Perkembangan Pembangunan Pengairan di Indonesia. First Part (2), *Direktorat Jenderal Pengairan, Departemen Pekerjaan Umum*, 6–7.

Wirosoemarto S. (1997b). Perkembangan Pembangunan Pengairan di Indonesia. First Part (1), *Direktorat Jenderal Pengairan, Departemen Pekerjaan Umum*, 132–133.

Young R. A. (2005). *Determining the Economic Value of Water: Concepts and Methods*. Resources for the Future. Washington D.C, United States of America.

Table 8.5 Timeline comparison of the governing system, water governance milestones and water engineering feats in Indonesia (1816-present).

Governing System	Water Governance Milestones	Water Engineering Feats
Old Dutch Colonial System (1816–1885) 1830- 1870 Enforced agriculture of main hard crops (*cultuurstelsel*) like coffee, quinine, tea and sugar cane	1830 Introduction of the enforced agriculture system under the Governor General of the Dutch Indies: van den Bosch 1850 Large-scale famine in the Demak and Semarang area, Central Java 1854 Establishment of the Water Branch under the Colonial Government (*Binnenlands Bestuur*) 1866 Establishment of the Directorate of Public Works (*Openbare Werkingen*)	Construction of the Sampean Baru Weir in Pasuruan, East Java 1830 Construction of the weir with modern engineering principles 1871 Further construction and development of the weir for irrigation purposes 1887 Extension of the irrigated area of the *prisé d'eau* of Sampean Baru.
Colonial System Transition (1885–1920) 1870 Trade liberalization	Initiation of water technocracy 1880 Establishment of the Agricultural Office 1885 Establishment of the Irrigation Brigade 1889 Development of the Irrigation Section 1890 Development of the overall irrigation plan of Java 1897 Establishment of the quality assurance commission (*Rentabiliteitcommissie*)	Pemali Irrigation Development 1889 Survey and investigation (under the engineer van Kol) 1893 Construction of the main embankment 1896 Irrigation development (under the engineer Lamminga) 1905 Improvement of the existing irrigation area 1910 Completion of the improved irrigation area
1902 Ethical politics enacted by the colonial Dutch Indies government 1907 Irrigation became one of the three important stakes of ethical politics. 1917 Establishment of the Dutch Indies people's council (*volksraad*)	1918 Huge flood (October 6) in Jakarta	Solo Valleiwerken 1881 Planning (Pierson) 1893 Construction 1898 Penghentian pekerjaan 1917 Prijetan Reservoir

Continued

Table 8.5 Timeline comparison of the governing system, water governance milestones and water engineering feats in Indonesia (1816-present) (*Continued*).

Governing System	Water Governance Milestones	Water Engineering Feats
Modern Colonialism (1920–1942)	Engineering deconcentration	Jakarta Flood Control Plan
1925- 1929 Decentralization of the colonial Dutch Indies Government 1929 Establishment of the West Java Province 1931 Establishment of the Central and East Java Provinces	1920 Separation of the water, roadways, haven and housing sections within the Director General of Public Works 1923 Establishment of the agricultural and farming centers 1928 Establishment of the water research station in Bandung 1929 Establishment of the Provincial Water Resources Office (*Provinciale Waterstaat*) 1936 Enactment of the Dutch Indies Water Law (*Algemeen Waterreglement*)	1922 Completion of the Western Flood Canal (under the engineer van Breen)Tangerang Irrigation Development 1916 Planning 1926 Construction 1936 Completion of the irrigation works
Transition Toward an Independent Indonesian State (1942–1949)	1949 Welfare plan (developed by the engineer J.W van Bloemmenstein)	None, due to the war and political circumstances.
Indonesian Parliamentary Democracy (1949–1959)	1957 Initiation of the war reparation scheme under the Japanese Government funding	1957 Musi River Improvement Project in Sumatera commences under Japanese Government funding 1958 Completion of the Southern Tulungagung Tunnel (Neyama) in the Brantas River Basin, East Java, under Japanese Government funding 1960 Initiation of the Jatiluhur Reservoir with French assistance (welfare plan from van Bloemmestein)

Continued

Table 8.5 Timeline comparison of the governing system, water governance milestones and water engineering feats in Indonesia (1816-present) (*Continued*).

Governing System	Water Governance Milestones	Water Engineering Feats
Indonesian Guided Democracy (1959–1966) 1959 Dissolution of the Constitution Body 1956 Nationalization of various foreign investments in Indonesia (Benteng Program) 1960 Establishment of the Conference of New Emerging Forces (CONEFO) as contender to the Old Forces (OLDEFO)	1959 Department of Public Works 1961 Establishment of project units under the Directorate General of Water Resources 1964 Reconstruction of the Department of Public Works and Electricity (Dwikora I Ministerial Cabinet) 1965 On force command applied to the project units	1959 The Brantas River Basin Plan assisted by Nippon Koei K.K., develops the first "modern" river basin master plan 1961 Construction of the Riam Kanan Reservoir 1962 Construction of the Karangkates Reservoir in the Brantas River Basin 1964 Construction of Selorejo Reservoir at the Konto River 1966 The big flood of the Bengawan Solo River (on March 6)
Indonesian New Order (1966–1998) 1967 Soeharto elected as President 1969 Indonesia re-enters its membership to the United Nations 1973 Development of international donor group through the International Governmental Group on Indonesia (IGGI)	1968 Establishment of the public works technocracy (Department of Public Works in the First Development Ministerial Cabinet) 1974 Law No 11 on Water Resources	1973 First basin-wide master plan for the Bengawan Solo River Basin in Central and East Java 1976 Establishment of the Water Resources Department at Brawijaya University 1978 First basin-wide master plan for the Jragung, Tuntang, Serang, Lusi and Juana River Basin in Central Java
	1980 Small-scale farmers credit initiative and large-scale irrigation plan under the Presidential Decree for food self-sustaining objective 1991 Reform in the Directorate General for Water Resources under the Department of Public Works	1981 First revision of the Jakarta flood control plan 1982 Completion of the Asahan Reservoir and the first stage of the hydro-electric power plant 1990 Establishment of Jasa Tirta I Public Corporation as a pioneering river basin organization (RBO) 1992 One million hectare swamp reclamation in Southern Borneo

Continued

Table 8.5 Timeline comparison of the governing system, water governance milestones and water engineering feats in Indonesia (1816-present).

Governing System	Water Governance Milestones	Water Engineering Feats
Reformed Bureaucracy (1998-sekarang) 1998 Indonesian reform action topples the New Order Regime 1999 Abdurrahman Wachid elected as President 2001 Megawati Soekarnoputri elected as President 2004 Susilo Bambang Yudhoyono elected as President	1998 Dissolution of the Department of Public Works, and formation of the Human Settlement and Infrastructure Department 2000 Trans-genetic seeds are introduced by multinational companies 2001 Re-establishment of the Public Works Department 2004 Law No 7 on Water Resources	1999 Government Regulation issued on Jasa Tirta I and Jasa Tirta II Public Corporation 2007 Initiation of the Jatigede using foreign investment bonds (China commercial bank) 2008 The big flood of the Bengawan Solo River (January 1) initiating the comprehensive flood control plan 2009 Initiation of the Jakarta Flood Plan (Eastern Flood Canal)

9 Partial authority allocation of regional water supply system in Indonesia and economic efficiency

Masamitsu Onishi[1], *Mangapul L. Nababan*[2], *Kiyoshi Kobayashi*[3]

[1] Graduate School of Engineering, Kyoto University, Japan. E-mail: onishi.masamitsu.7e@kyoto-u.ac.jp

[2] Center for Strategic Studies, Ministry of Public Works of the Republic of Indonesia. E-mail: mangapul_nababan@pu.go.id

[3] Graduate School of Management, Kyoto University, Japan. E-mail: kobayashi.kiyoshi.6n@kyoto-u.ac.jp

Abstract

Our study is motivated by a problem that we observed in the Bandung region of Indonesia, where the operation of the water supply system has been affected by the decentralization policy of 2001. The decentralization promoted by the government of Indonesia has vested in local governments the authority to determine substantial issues related to the water supply system. As is observed by Nababan *et al.* (2012), a case in the Bandung region where the authority to operate a water supply system covering multiple local governments' territories was monopolized by one local government among them. In this study, we are interested in exploring, from an economic point of view, the problematic consequences of this 'partial authority allocation.' Our study aims to develop a descriptive model to explain the mechanism of how economic inefficiency arises from partial authority allocation. In addition, we discuss feasible policies to overcome the identified problem.

Keywords: Regional water supply system, decentralization, partial authority allocation

9.1 INTRODUCTION

Without water, no one can survive. Water is definitely essential for life. But water is not ubiquitous. Humans have sought to secure water availability throughout the ages. To achieve stable availability of clean water is still a great challenge in developing countries. The millennium development goals of UNDP (2003) include halving by 2015 the proportion of people without sustainable access to safe drinking water and basic sanitation. A number of reports have pointed out the inadequacy of water supply in urban areas, where poor households seldom have networked water supply access (UNCHS, 2003 and 2006; UNDP, 2006; UNWWAP, 2003 and 2006).

Allocation of water resources in a region is usually controlled under governmental administration, as it is subject to the problem of common pool resources. In addition, a water supply system requires extensive infrastructure; private ownership would lead to an inefficient level of water supply due to the monopolistic power often associated with such ownership. The question arises, then, of which governmental level should hold authority over a regional water supply system.

According to the famous principle of fiscal equivalence developed by Olson (1969), we have to have one layer of the government for each public good, although this does not take into account individuals' mobility. As water is regional in the sense that it is not easily mobile, it is usually administrated by the local government. However, as the territories of a local government's jurisdictions are formed as a result of historical contingencies, the regional territory of a unit of water supply system could involve multiple jurisdictions. If a unit of water supply system involves multiple jurisdictions, it is common for an independent organization to administrate the unit of water supply system under the control of involved local governments.

If multiple local governments are involved in the control of the organization, the allocation of authority over it matters. If authority over the organization operating on behalf of multiple jurisdictions is partially allocated to a specific local government among them, the organization might not have appropriate incentive to maximize the welfare of all jurisdictions covered. This study reveals that inefficiency in water provision due to the partial allocation of authority over a water supply system which covers multiple jurisdictions will become more serious when the availability of water resources fluctuates seasonally. More specifically, without the seasonal fluctuation in the availability of water resources, non-discriminatory pricing within the territory of the water supply system will guarantee efficient water supply even if authority over the water supply system is partially allocated. However, if the seasonal fluctuation in the availability of water resources exists, non-discriminatory pricing is not sufficient to guarantee efficiency in supplying scarce water if authority is partially allocated.

Our study is motivated by a problem that we observed in the Bandung region of Indonesia, where the operation of the water supply system has been affected by the decentralization policy of 2001. The decentralization promoted by the government of Indonesia has vested in local governments the authority to determine substantial issues related to the water supply system. Unfortunately, the actual implementation of the decentralization policy in Indonesia has not necessarily regarded the need to coordinate local governments' conflicting interests. Nababan *et al.* (2012) observed a case in the Bandung region where the authority to operate a water supply system covering multiple local governments' territories was monopolized by one local government among them. In this study, we are interested in exploring, from an economic point of view, the problematic consequences of this 'partial authority allocation.' Our study aims to develop a descriptive model to explain the mechanism of how economic inefficiency arises from partial authority allocation. In addition, we discuss feasible policies to overcome the identified problem.

The paper is organized as follows. Section 9.2 describes the current institutional environment in the Bandung region after decentralization in Indonesia, which has motivated us to conduct this study. In section 9.3 and 9.4, a descriptive model is developed to analyze economic efficiency in the current institutional environment. Section 9.3 analyzes the case of symmetric local governments, whereas section 9.4 analyzes the case of asymmetric local governments. As a result, we point out that asymmetricity regarding the availability of water is a source of potential conflict of interest and leads to economic inefficiency. In section 9.5, several feasible policies are proposed based on the implications of our analysis. Section 9.6 concludes the paper.

9.2 BASIC IDEA OF THE STUDY

9.2.1 Institutional System of Water Supply System After Decentralization in the Bandung Region

Our study is motivated by problems related to the institutional consequences of decentralization in Indonesia in 2001, as observed in the Bandung region. In Indonesia, the utility of water supply is provided by public enterprises called PDAM (Perusahaan Daerah Air Minum). PDAM is responsible for seeking and developing water sources and facilities for water treatment and transmission. Since the 2001 decentralization, PDAM has consisted completely of government-subsidized entities controlled by local governments. The territory covered by a single PDAM is not necessarily a single local government. Rather, for geographical reasons, it is common for a PDAM to cover multiple local governments' territories.

This is the case in the Bandung region. PDAM Tirta Raharja supplies water to Bandung Regency, West Bandung Regency and Cimahi City. Any decision making of PDAM Tirta Raharja affects the concerns of the municipalities covered. However, ownership of PDAM Tirta Raharja has been allocated only to Bandung Regency since decentralization was promoted, with West Bandung Regency and Cimahi city excluded from the formal decision-making process. Cimahi City is recognized as originally being a military area where a number of military facilities are located, but it has recently experienced population growth leading to increased demand for water for drinking and household use, as well as for commercial use. On the other hand, Bandung Regency, which holds 100% ownership of PDAM Tirta Raharja, is rather sparsely populated and covers a mountainous area. These conditions imply that, compared to people in Cimahi City, it is easier for people in Bandung Regency to acquire alternative water sources other than PDAM.

9.2.2 Problem Identification

This paper is partly motivated by feelings of frustration among Cimahi City government officials. Due to the increase in population, they wish to expand the networked water supply system. However, the Cimahi City government has no control over PDAM Tirta Raharja. It can be stated intuitively that investment in a pipeline network for Cimahi City is inadequate. Our study aims to develop a mechanism to explain, from a theoretical point of view, underinvestment in the pipeline network in the populated area under the current governance system of PDAM Tirta Raharja.

We focus upon the asymmetricity in geographical conditions between Cimahi City and Bandung Regency, which holds authority over PDAM Tirta Raharja. Cimahi City is steadily growing, but alternative water resources are not rich compared to those of the surrounding mountainous areas, including Bandung Regency. It is shown that the difference in richness of alternative water sources could be a critical factor causing underinvestment in Cimahi City. In addition, PDAM Tirta Raharja gains a positive surplus from Cimahi City, which is transferred for use as investment in expanding the network in Bandung Regency, when the government of Bandung Regency behaves rationally.

Another source of the problem of partial authority allocation is rather a technical attribute. Pipeline is essential for connecting to the water supply network. Without pipeline, people cannot use the networked water. This implies that investment in pipeline is economically interpreted as purchasing a financial 'option' to use the networked water. The value of the option apparently depends on the availability of alternative water sources. A household with fewer alternative water sources may value the option, i.e. pipeline, more highly than would a household with more alternative sources.

In summary, the purpose of this study is to prove theoretically that partial authority allocation could be problematic if there exists a difference in richness of alternative water sources and pipeline as an option.

9.3 MODEL

9.3.1 Setting

The economy consists of two regions, Region 1 and Region 2. The population of the two regions is symmetric and standardized as 1. Individuals of the economy are assumed to be immobile. PDAM is a corporation established to supply water. It is assumed that only the Region 1 government has authority to determine the price of water and investment for pipelines in the two regions. This assumption reflects the partial ownership allocation of water supply system in Indonesia. The price of PDAM water cannot be discriminated between Region 1 and Region 2.

9.3.2 Technology

Costs of PDAM water supply system consist of two components: cost associated with the volume of supplied water (e.g., water treatment cost) and cost associated with the initial investment which determines the capacity of water supply (e.g., pipeline construction). Fixed cost components are not considered. The total cost of PDAM water supply system is formulated as

$$C(x_1, x_2, n_1, n_2) = c(x_1 + x_2) + c_f(n_1 + n_2) \tag{9.1}$$

x_i denotes the aggregate volume of supplied water for Region i and n_i denotes the number of households that have access to the pipeline network of PDAM. Hence, n_i is the maximum volume of water supply (i.e. capacity of water supply system) in Region i, i.e. $x_i \leq n_i$. n_i is called hereinafter 'network capacity'. A fixed cost of the investment for facilities such as water treatment plants that benefit both Region 1 and Region 2 is not considered.

For simplicity, in the following analysis,

$$c + c_f = 1 \tag{9.2}$$

$$c_f = kc, \ (k > 0) \tag{9.3}$$

is assumed.

9.3.3 Demand

Households consume one unit of water. Households do not necessarily use water supplied by PDAM exclusively. It is very common for the Indonesian to use private wells or a community-based water supply system, hereinafter called alternative water sources. Hence, households choose PDAM or alternative water sources in consuming one unit of water. ρ is the cost of using alternative water sources, which is distributed according to the cumulative distribution function $F_i(\rho) = a_i\rho$ in Region i, where $\rho \in [0, 1/a_i]$. a_i is a parameter which represents the cost degree of acquiring alternative water sources. The larger a_i means the relative easiness of acquiring alternative water sources.

Given the unit price of PDAM water p, a household whose cost of using alternative water sources is ρ chooses PDAM if $p \le \rho$ and alternative water sources if $p > \rho$. The aggregate demand function of PDAM water without the network capacity constraint, which is called the potential aggregate demand hereinafter, is calculated as

$$\bar{x}_i(p) = 1 - F_i(p) = 1 - a_i p \tag{9.4}$$

The inverse aggregate demand function is derived as

$$\bar{P}_i(x_i) = \frac{1-x_i}{a_i}. \tag{9.5}$$

The symbol "‾" denotes that demand is 'potential' and unconstrained by the network capacity. When the network capacity constraint is considered, a household may not obtain PDAM water even if the price of PDAM water is lower than the cost of alternative water sources. If the potential aggregate demand exceeds the network capacity, i.e. $\bar{x}_i(p) > n_i$, the network capacity constraint is effective. $\bar{x}_i(p) > n_i$ implies $\bar{P}_i(n_i) > p$. In this case, it is assumed that households with $\rho \ge \bar{P}_i(n_i) = (1-n_i)/a_i$ consume one unit of PDAM water. And households with $\rho < \bar{P}_i(n_i)$ use alternative water sources. Hence, the 'real' aggregate demand function with the capacity constraint given p is

$$x_i(p, n_i) = \begin{cases} 1 - a_i p & \text{if } 1 - a_i p < n_i \\ n_i & \text{if } 1 - a_i p \ge n_i \end{cases} \tag{9.6}$$

$\bar{P}_i(n_i)$ is a threshold to determine whether the capacity is fully utilized or not depending on the setting of p. In other words, in order to fully utilize the network capacity n_i, the price must be set at less than $\bar{P}_i(n_i)$.

In addition,

$$\bar{x}_i(c + c_f) = 1 - a_i(c + c_f) \gg 0 \text{ for } i = 1, 2 \tag{9.7}$$

is assumed. This assumption guarantees that when the water is supplied at the price of minimum marginal cost, sufficient aggregate demand exists.

9.3.4 Cost of Water Acquisition

The total cost of water acquisition in Region i is defined as

$$TC_i = px_i(p, n_i) + \int_0^{\max[p, \bar{P}_i(n_i)]} \rho dF_i(\rho). \tag{9.8}$$

The first term of LHS denotes the cost incurred by households that use PDAM water, and the second term denotes the cost for usage of alternative water sources.

PDAM gains revenue from households using PDAM water as compensation for being responsible for water supply and network investment. The revenue of PDAM coming from Region *i* is represented by $R_i = px_i(p,n_i)$ as PDAM cannot differentiate the price in Region 1 and Region 2. The cost of supplying PDAM water for Region *i* is

$$C_i = cx_i(p,n_i) + c_f n_i. \tag{9.9}$$

For the analytical convenience, define

$$\theta_i = \frac{x_i(p,n_i)}{n_i} \tag{9.10}$$

as denoting the ratio of usage to the capacity of pipeline. θ_i must be $0 \leq \theta_i \leq 1$. The cost of PDAM water supply Region *i*, C_i, is written as

$$C_i = m_i n_i \text{ and} \tag{9.11}$$

$$m_i = \frac{\sigma_i + k}{1+k}, \tag{9.12}$$

where $\sigma_i = 1/\theta_i (\geq 1)$. m_i is regarded as the apparent marginal cost of water supply in Region *i*. Profit of PDAM is written as

$$\pi_i = R_i - C_i = (p - m_i)n_i \tag{9.13}$$

When the capacity of pipeline is fully utilized, the marginal cost is 1. However, if the capacity of pipeline is not fully utilized, the marginal cost becomes larger than 1 as θ_i decreases.

The profit of PDAM is the sum of the profit from Region 1 and Region 2. That is,

$$\pi = \pi_1 + \pi_2 \tag{9.14}$$

The followings are some important properties of the total cost of water acquisition, which are frequently used in the following analysis.

$$\frac{\partial TC_i}{\partial p} > 0 \tag{9.15}$$

$$\frac{\partial TC_i}{\partial n_i} \begin{cases} < 0 & \text{if } p < \overline{P}_i(n_i) \\ = 0 & \text{if } p \geq \overline{P}_i(n_i) \end{cases} \tag{9.16}$$

(9.15) means that the total cost of water acquisition is always strictly increasing in the price of PDAM water. (9.16) means that expanding the supplying capacity contributes to decreasing the total cost when the capacity is constrained, but not when this is not the case.

9.3.5 The Case of Impartial Authority Allocation

When the authority over the PDAM is impartially allocated, it concerns the welfare of Region 1 as well as Region 2. The price of PDAM water and the level of network capacity must be set to minimize the total cost of water acquisition in Region 1 and Region 2, which is formulated as follows:

$$\min_{p,n_1,n_2} TC_1 + TC_2 \tag{9.17}$$

subject to

$$\pi_1 + \pi_2 = 0. \tag{9.18}$$

Solving the above problem, the following **Lemma 1** is obtained.

Lemma 1: When the authority over the PDAM is impartially allocated, the unit price of water and network capacity in Region 1 and Region 2 are set as follows (see the Supplementary Information for the proof).

$$p^* = 1$$

$$n_1^* = 1 - a_1$$

$$n_2^* = 1 - a_2$$

In the process of proving **Lemma 1**, it is shown that the supplying capacities should be set to satisfy exactly the potential aggregate demand in each region. Redundant supplying capacity is always undesirable. Secondly, the zero profit constraint of PDAM requires the price to coincide with the marginal cost of water supply, including two components of costs, i.e. c and c_f. In summary, when the authority over the PDAM is impartially allocated, the price of PDAM water is set at the level equivalent to the marginal cost of water supply. In addition, the supplying capacities in both regions cover the potential aggregate demand exactly.

9.3.6 The Case of Partial Authority Allocation

This subsection is for analysis of the case when the authority over the PDAM is partially allocated to Region 1, and Region 2 has no influence on the PDAM's decision. The government of Region 1 is benevolent in the sense that it represents the welfare of people exclusively in Region 1 and attempts to minimize the aggregate cost of households to acquire water in Region 1. PDAM is in principle a public corporation. Hence, price is set at the level of cost recovery basis; i.e., the profit of PDAM must be zero.

Therefore, the price of PDAM water and network capacity in Region 1 and Region 2 are effectively set by the government of Region 1. The problem of the government in Region 1 is formulated as minimization of the total cost of water acquisition in Region 1.

$$\min_{p,n_1,n_2} TC_1 \quad \text{subject to (9.18).} \tag{9.19}$$

By solving the above optimization problem, the following **Proposition 1** is derived (see the Supplementary Information for the proof).

Proposition 1: The government in Region 1 sets the price of PDAM water $p^\circ = 1$ and $n_1 = 1 - a_1$. Decision on n_2 has no influence on the welfare of Region 1 as long as it is $n_2 \in [0, 1 - a_2]$.

Non-discriminatory pricing will provide the authoritative government an incentive to set the price at the efficient level under the above setting. For the PDAM working for the welfare of Region 1, the capacity level in Region 2 does not matter in the sense that it does not affect the welfare of Region 1. Therefore, claiming the capacity level $n_2 = 1 - a_2$ by the Region 2 government does not create any conflict with the objective of the PDAM with partial authority allocation. However, the next section shows that non-discriminatory pricing does not work to guarantee efficiency if the availability of alternative water sources fluctuates seasonally.

9.4 FLUCTUATING AVAILABILITY OF ALTERNATIVE WATER SOURCES

9.4.1 Assumption

The cost of using alternative water sources fluctuates depending on the season. In fact, private wells and springs are more prone to shortages than river water, which is the main source of PDAM. The previous section identified the principle that the supplying capacity must exactly fulfill the potential demand of PDAM water. However, the supplying capacity is not variable depending on seasons. Therefore, the fluctuation of the potential demand for PDAM water makes it impossible to apply this '*fulfilling*' principle. Hence, the supplying capacity will be redundant in some seasons, whereas it may be a strictly binding constraint in other seasons.

The aggregate demand for PDAM water in Region 1 is assumed to be

$$\begin{cases} \bar{x}_1^N(p) = 1 - a_1^N p & \text{normal season} \\ \bar{x}_1^D(p) = 1 - a_1^D p & \text{dry season} \end{cases} \tag{9.20}$$

i.e., $F_1^j(p) = a_1^j p$ $(j = B,D)$ where $a_1^N > a_1^D$. For simplicity of analysis, assume $a_1^D = a_2 = a$ and $a_1^N = \delta a$ where $\delta > 1$. The length of the dry season is denoted by γ and of the normal season $1-\gamma$ where $0 \le \gamma \le 1$. The price of PDAM water is constant through a whole year once it is determined. The government of Region 1 decides the price of PDAM water and the water supply capacity in the two regions to minimize the total cost incurred for water usage.

The inverse demand function is also defined in the same manner as above.

$$\bar{P}_1^j(x_i) = \frac{1-x_i}{a_1^j} \quad \text{for } j = N,D \tag{9.21}$$

$a_1^N > a_1^D$ implies $\bar{P}_1^N(n_1) < \bar{P}_1^D(n_1)$. Following the definition of 'real' aggregate, the demand function in Region 1 is represented as

$$x_1^j(p,n_1) = \begin{cases} 1 - a_1^j p & \text{if } 1 - a_1^j p < n_1 \\ n_1 & \text{if } 1 - a_1^j p \ge n_1 \end{cases} \quad \text{for } j = N,D. \tag{9.22}$$

In addition, the following condition is assumed.

$$1 - a_1^N(c + c_f) \ge 0 \tag{9.23}$$

9.4.2 The Cost of Water Acquisition and the Profit of PDAM

The total cost of water acquisition in Region 1 fluctuates with the seasonal change in the availability of alternative water sources. Then the total cost of water acquisition in Region 1 in each season is represented by

$$TC_1^j = px_1^j\left(p,n_1\right)+\int_0^{\max\left[p,\,\bar{P}_1^j(n_1)\right]}\rho dF_1^j\left(\rho\right),\text{ for } j=N,D. \tag{9.24}$$

Therefore, the total cost of water acquisition in Region 1 through a year is calculated as

$$TC_1 = \gamma TC_1^D + (1-\gamma)TC_1^N. \tag{9.25}$$

On the other hand, the season-dependent cost for supplying PDAM water in Region 1 is

$$C_1^j = cx_1^j\left(p,n_1\right)+c_f,\text{ for } j=N,D. \tag{9.26}$$

Therefore, the cost for PDAM water through a year is

$$C_1 = \gamma C_1^L + \left(1-\gamma\right)C_1^H,\text{ for } j=N,D. \tag{9.27}$$

For analytical convenience, define

$$\theta_1^j = \frac{x_1^j\left(p,n_1\right)}{n_1} \tag{9.28}$$

as denoting the ratio of usage to the capacity of pipeline. θ_1^j must be $0 \le \theta_1^j \le 1$. The cost of PDAM water supply C_i is written as

$$C_1 = m_1 n_1 \text{ and} \tag{9.29}$$

$$m_1 = \gamma m_1^D + (1-\gamma)m_1^N \tag{9.30}$$

$$m_1^j = \frac{\sigma_1^j + k}{1+k}, \tag{9.31}$$

where $\sigma_1^j = 1/\theta_1^j(\ge 1)$. m_1^j is regarded as the apparent marginal cost of water supply in Region i in season j.

For the analytical convenience, define

$$M = 1 + \mu(m_1 - 1) \text{ and} \tag{9.32}$$

$$\mu = \frac{n_1}{n_1 + n_2}. \tag{9.33}$$

Using the above notation, the total profit of PDAM is written as

$$\pi = (p - M)(n_1 + n_2). \tag{9.34}$$

M is regarded as the apparent marginal cost of water supply in total. *M* is dependent on the price, the supplying capacity and the ratio of seasonal length.

9.4.3 The Case of Impartial Authority

By the same token as in section 9.3, the case of impartial authority requires the PDAM to minimize the sum of the total cost of water acquisition in Region 1 and Region 2. Hence the optimization problem is formulated as

$$\min_{p, n_1, n_2} TC_1 + TC_2 \tag{9.35}$$

subject to

$$\pi = 0. \tag{9.36}$$

For simplicity in the analysis, assume $\delta a \approx 0$ and

$$\frac{1-\gamma}{\gamma} < (1+k)(\delta - 1). \tag{9.37}$$

Solving the above optimization problem, the following **Lemma 2** can be obtained (see the Supplementary Information at the end of this chapter for the proof).

Lemma 2: If the authority over PDAM is impartially allocated between Region 1 and Region 2, the price and the supplying capacity in each region are set as follows.

$$p^{**} > 1$$

$$1 - \delta a p^{**} < n_1^{**} \leq 1 - a p^{**}$$

$$n_2^{**} = 1 - a p^{**}$$

(9.37) is an assumption to guarantee that deviating the price from the marginal cost and extending the supplying capacity in Region 1 contribute to decreasing the total cost. If PDAM supplies water at a price equivalent to the marginal cost, due to the zero profit constraint, PDAM water is insufficiently supplied in the dry season due to the shortage of pipelines. In order to compensate for the shortage of pipelines in the dry season in Region 1, price has to be increased to meet the balance of PDAM accounting. On the other hand, an increase in price reduces the usage of water in the normal season in Region 1 as well as in Region 2. In addition, expanded network is redundant in the normal season. Therefore, some households in Region 1 do not use PDAM water in the normal season; hence there is no revenue from them to PDAM in the normal season. In fact, the profit coming from each region is

$$\pi_1 = (p - m_1) n_1 < 0 \tag{9.38}$$

$$\pi_2 = (p - m_2) n_2 = (p - 1) n_2 > 0. \tag{9.39}$$

This means that Region 2 pays more than the marginal cost, whereas Region 1 pays less. It is interpreted that the surplus is transferred from Region 2 to Region 1 because of the redundancy in Region 1.

To what extent the surplus is transferred from Region 2 to Region 1 depends on the difference in availability of alternative water sources between the dry season and the normal season, and the length of the dry season.

9.4.4 The Case of Partial Authority Allocation

When the authority over the PDAM is partially allocated to the government of Region 1, the PDAM pursues minimizing the total cost of water acquisition in a year. Keeping the assumption (9.37),

$$\min_{p, n_1, n_2} TC_1 \tag{9.40}$$

subject to (9.36).

Solving the above problem indicates that the optimal choice of the Region 1 government is derived as shown in **Proposition 2** (see the Supplementary Information at the end of this chapter for the proof).

Proposition 2 With the seasonal fluctuation of the aggregate demand in Region 1, the price of PDAM water $p^{\circ\circ}$ and the supplying capacity in Region 1 and Region 2, $n_1^{\circ\circ}$, $n_2^{\circ\circ}$ satisfies

$$p^{\circ\circ} > p^{**}$$

$$n_1^{\circ\circ} > n_1^{\circ\circ}$$

$$n_2^{\circ\circ} = 1 - p^{\circ\circ} < n_2^{**}$$

Proposition 2 implies that the price under partial authority allocation is higher than that under impartial authority allocation. In addition, partial authority allocation allows PDAM to invest more for the network capacity in Region 1 whereas less for Region 2 than the case of impartial authority allocation. Partial authority allocation is preferable for Region 1, but not for Region 2. An increase in the price generates more surplus from Region 2. The surplus in Region 2 is spent on expanding the network capacity in Region 1, which enables a decrease in the cost of water acquisition in Region 1 in the dry season due to the supplying capacity constraint.

Comparison of the cases of partial authority allocation and impartial authority allocation tells us that the households in Region 2 are in a disadvantageous position. Firstly, Region 2 pays more than the cost spent for their PDAM water usage. Secondly, they could enjoy more usage of PDAM water if the price of PDAM water were set on a cost recovery basis.

Section 9.5 discusses the sources of this problem and gives suggestions for alternative policies to mitigate the problem.

9.5 DISCUSSIONS

9.5.1 Externalities

The problem of partial authority allocation between the regions is a matter of *externalities* arising from unique attributes of the case that we assume for depicting the institutional environment in the water supply system in the Bandung region. Fundamental assumptions set in the above analysis are summarized as follows:

1. The supplier of the networked water covers multiple jurisdictions for geological reasons and the profit of the water supplier is constrained to be zero.
2. There is no discrimination of water prices between the regions.
3. There is asymmetricity in the availability of alternative water sources between the regions.
4. There is seasonal fluctuation of the networked water source.

Assumption 1) is related to the issue of water rights and economy of scale technology. If each region has plenty of raw water sources and there is no economy of scale, there is no economic reason for a supplier of networked water to cover multiple jurisdictions. Although our model does not assume the economy of scale and water rights setting explicitly, the supplier's coverage of multiple jurisdictions can be justified by the technical assumption 1).

Assumption 2) is regarded as a political constraint rather than an economic constraint.

Assumption 3) and 4) are conditions necessary for generating a difference in the usage of supplying capacity between or among regions. The fluctuation regarding the availability of alternative water sources can create redundancy in the supplying capacity of pipeline. This is because pipelines cannot be adjusted to the volume of PDAM water usage. In this sense, an investment in pipelines is regarded as a fixed cost.

Our primary motivation for conducting this study is to identify the problem of partial authority allocation over the PDAM, which is the way in which the water supply in Bandung region is organized. To reprise the case there, PDAM Tirta Raharj covers areas of several jurisdictions including Bandung Regency, which holds monopolistic authority over PDAM, and Cimahi City. Region 1 in the model is regarded as Bandung Regency and Region 2 as Cimahi City.

This difference is typically apparent between a rural area (i.e. Bandung Regency) and an urban area (i.e. Cimahi City). Bandung Regency, with richer alternative water sources, may not use the full supplying capacity of pipeline in the non-dry season.

Partial allocation of authority over the PDAM with multiple jurisdictions generates externalities under the above settings. According to the analytical result in 9.3, the price set under the partial authority allocation is efficient and the supplying capacity in Cimahi City, with fewer alternative water sources, does not affect the welfare of the rural area. However, the analytical result in 1.4 indicates that introducing the fluctuating demand of PDAM water, hence the redundancy without the fluctuation of aggregate demand, is a source of externalities on Cimahi City.

Bandung Regency suffers less from price increases than Cimahi City, as the former has greater access to alternative water sources other than PDAM water. Rather, the interest of Bandung Regency is in expanding the supplying capacity, which is constrained in the dry season. Therefore, Bandung Regency is tempted to expand the supplying capacity in its territory by increasing the revenue of PDAM. The increase of PDAM water generates a surplus from Cimahi City, which can be utilized by Bandung Regency to expand the supplying capacity in its own territory.

9.5.2 Policy Implication

Externalities can be overcome by efficient negotiation, according to the Coasian theorem (Coase, 1960). This theorem tells us that any externalities can be internalized by efficient negotiation among stakeholders regardless of the initial allocation of ownerships or authorities. Therefore, according to the Coasian theorem, Cimahi City may be motivated to ask Bandung Regency to decrease the price of water and to expand the supplying capacity in Cimahi City, by offering a direct money transfer from Cimahi City to Bandung Regency.

However, the Coasian theorem is concerned only with economic efficiency and not with fairness. For Cimahi City to join a negotiation process with Bandung Regency would imply that Cimahi City accepted the current regime of partial authority allocation over the PDAM. But people in Cimahi City will refuse any such negotiation, so as to avoid appearing to accept the partial authority allocation as an accomplished fact.

The failure of efficient negotiation between Cimahi City and Bandung Regency implies the necessity of intervention by a higher-ranking authority, such as the provincial government or the national government. Although a governing system encompassing the provincial government is beyond the scope of the present study, the problem identified above will not be solved unless the provincial government becomes more aware of its role in coordinating the interests of local governments such as cities and regencies.

9.6 CONCLUSION

Partial authority allocation over the networked water facility system in Indonesia is a consequence of the decentralization policy of 2001. Our analysis shows that partial authority allocation can be problematic when regional asymmetricity regarding the availability of alternative water sources and seasonal fluctuation exist. The price of networked water is higher than the efficient level, and investment for pipelines in the urban area is insufficient, whereas in the rural area it is excessive. Hence, the insufficient penetration of PDAM water in the urban area can be partly attributed to the current regime of partial allocation of authority over the PDAM.

In addition, we have pointed out that, for political reasons, negotiation to solve the problem will not work. This kind of partiality problem must be solved by a higher rank of government, such as the provincial or central government.

9.7 ACKNOWLEDGEMENT

We thank Mr. Hery Antasari and staff, from Bapeda Cimahi City Local Government, and Ms. Hendriati from PDAM Tirta Raharja. We are also very thankful to Dr. Ibnu Syabri, and his colleagues in School of Architecture, Planning and Policy Development ITB, Bandung.

9.8 REFERENCES

UNDP (2003). *Millennium Development Goals: A Compact for Nations to End Human Poverty*, United Nations Development Program, New York.

UNCHS (2003). *Local Action for Global Goals: Water and Sanitation in the World's Cities*, Earthscan, London.

UNCHS (2006). *Meeting Development Goals in Small Urban Centres: Water and Sanitation in the World's Cities 2006*, Earthscan, London.

UNDP (2006). Beyond scarcity: power, poverty, and the global water crisis. *Human Development Report 2006*, United Nation Development Program, New York.

UNWWAP (2003). *Water for People, Water for Life*, World Water Assessment Program, FAO, Rome.

UNWWAP (2006). *Water: A Shared Responsibility*, United Nations World Assessment Program, Rome.

Olson M. (1969). The principle of "fiscal equivalence": The division of responsibilities among different levels of government. *American Economic Review*, **59**, 479–487.

Nababan M. L., Onishi M., Kobayashi K. (2012). *Asymmetric ownership between municipalities in water supply system provision in Indonesia*, mimeo. The 3rd International Workshop on Water Supply Management System and Social Capital, Brawijaya University, Makassar, Indonesia.

Coase R. H. (1960). The problem of social cost. *Journal of Law and Economics*, **3**, 1–44.

SUPPLEMENTARY INFORMATION

PROOF OF PROPOSITIONS

(Proof of Lemma 1)

Firstly, consider a point (p, n_1, n_2) that satisfies (9.18) and $p > \bar{P}_i(n_i)$, i.e. the capacity is not constrained. In this case, cutting the supplying capacity does not change the total cost of water acquisition. Cutting the supplying capacity generates a surplus for PDAM which enables it to decrease the price of water and to use the redundant supplying capacity since (9.15) holds. Therefore, any (p, n_1, n_2) which satisfies $p > \bar{P}_i(n_i)$ cannot be optimal.

Secondly, a point (p, n_1, n_2) that satisfies (9.18) and $p \le \bar{P}_i(n_i)$ or $m_i = 1$, i.e. the capacity is strictly constrained. The zero profit constraint of PDAM (9.18) is written as

$$(p-1)(n_1+n_2)=0. \tag{9.41}$$

When $n_1 + n_2 \ge 0$, the zero profit constraint requires $p = 1$.

Then the problem is reformulated as the problem of optimizing the supplying capacity n_1 and n_2 for minimizing the total cost of water acquisition given $p = 1$ and $p \le \bar{P}_i(n_i)$, where zero profit constraint always holds.

If $\bar{P}_i(n_i) \ge 1$, the associated change of the total cost of water acquisition due to the small change in the supplying capacity is derived as

$$\begin{aligned} &d(TC_1+TC_2) \\ &=\{1-\bar{P}_1(n_1)\}dn_1+\{1-\bar{P}_2(n_2)\}dn_2. \end{aligned} \tag{9.42}$$

It is apparent that as long as $\bar{P}_1(n_1) \ge 1$ and $\bar{P}_2(n_2) \ge 1$, the small change in the supplying capacity contributes to reducing the total cost of water acquisition. In addition, $\bar{P}_i(n_i)$ is monotonically decreasing in the supplying capacity n_i. Therefore, the total cost of water acquisition is minimized when $\bar{P}_1(n_1) = 1$ and $\bar{P}_2(n_2) = 1$ holds.

(Proof of Proposition 1)

The only difference between the problem of partial authority allocation and the impartial authority allocation is the objective function. The logic of proof is fundamentally the same as the one for **Lemma 1**.

$$dTC_1 = \{1-\bar{P}_1(n_1)\}dn_1 \tag{9.43}$$

Minimizing the objective function, i.e. the total cost of water acquisition in Region 1, requires $\bar{P}_1(n_1) = 1$. However, the supplying capacit.y in Region 2 n_2 is not relevant to the objective of PDAM controlled by the government in Region 1.

(Proof of Lemma 2)

Firstly, consider a point (p, n_1, n_2) that satisfies (9.36) and $p > \bar{P}_1^D(n_1)$ and $p > \bar{P}_2(n_2)$, i.e. the capacity is not constrained in any season. In this case, cutting the supplying capacity does not change the total cost of water acquisition. Cutting the supplying capacity generates a surplus for PDAM which enables it to decrease the price of water and to use the redundant supplying capacity since (9.15) holds. Therefore, any (p, n_1, n_2) which satisfies $p > \bar{P}_1^D(n_1)$ and $p > \bar{P}_2(n_2)$ cannot be optimal.

Secondly, a point (p, n_1, n_2) that satisfies (9.36) and $p \le \bar{P}_1^N(n_1)$ and $p \le \bar{P}_2(n_2)$, i.e. the supplying capacity, is constrained in any season. In this case, $m_1 = m_2 = 1$ holds and the zero profit constraint (9.36) is written as

$$(p-1)(n_1+n_2)=0 \qquad (9.44)$$

When $n_1 + n_2 \ge 0$, the zero profit constraint requires $p = 1$. Following the same logic used in the proof of **Lemma 1**, $\bar{P}_1^N(n_1)=1$ and $\bar{P}_2(n_2)=1$ holds.

Thirdly, a point (p, n_1, n_2) that satisfies (9.36) and $\bar{P}_1^N(n_1) \le p \le \bar{P}_1^D(n_1)$ and $p \le \bar{P}_2(n_2)$, i.e. the supplying capacity, is constrained in the dry season, but redundant in the normal season. In this case, $m_1 \le 1$ holds.

$$\pi = (p-M)(n_1+n_2). \qquad (9.45)$$

When $n_1 + n_2 \ge 0$, the zero profit constraint requires $p = M$. Contrary to the case of $p \le \bar{P}_1^N(n_1)$ and $p \le \bar{P}_2(n_2)$, expanding the supplying capacity is accompanied by the associated change in the apparent marginal cost M. Note the $m_1^D = 1$ and $m_1^D \le 1$. The associated change in M due to the small change in supplying capacity n_1 is

$$\frac{\partial M}{\partial n_1} = \frac{\partial \mu}{\partial n_1}(m_1 - 1) + \mu \frac{\partial m_1}{\partial n_1}$$

$$\frac{\partial \mu}{\partial n_1} = \frac{n_2}{(n_1+n_2)^2}(>0)$$

$$\frac{\partial m_1}{\partial n_1} = (1-\gamma)\frac{\partial m_1^N}{\partial n_1}$$

$$\frac{\partial m_1^N}{\partial n_1} = \frac{1}{1+k}\frac{\partial \sigma_1^N}{\partial n_1}$$

$$\frac{\partial \sigma_1^N}{\partial n_1} = \frac{1}{x_1^N(p,n_1)} > 0$$

Hence a small change in the supplying capacity always induces an increase in the apparent marginal cost .

$$\frac{\partial M}{\partial n_2} = \frac{\partial \mu}{\partial n_2}(m_1 - 1)$$

$$\frac{\partial \mu}{\partial n_2} = -\frac{n_1}{(n_1+n_2)^2}(<0)$$

A small change in the price also induces a change in the apparent marginal cost.

$$\frac{\partial M}{\partial p} = \mu \frac{\partial m_1}{\partial p} \tag{9.46}$$

$$\frac{\partial m_1}{\partial p} = (1-\gamma)\frac{\partial m_1^N}{\partial p} \tag{9.47}$$

$$\frac{\partial m_1^N}{\partial p} = \frac{1}{1+k}\frac{\partial \sigma_1^N}{\partial p} \tag{9.48}$$

$$\frac{\partial \sigma_1^N}{\partial p} = \frac{a_1^N}{\{x_1^N(p,n_1)\}^2} > 0 \tag{9.49}$$

Hence a small change in the price always induces an increase in apparent marginal cost M.

In addition, the second-order of the marginal change of M due to the small change of p is

$$\frac{\partial^2 M}{\partial p^2} > 0. \tag{9.50}$$

As the price increases, the associated change of M becomes larger.

Consider the objective function. When $\bar{P}_1^N(n_1) \le p \le \bar{P}_1^D(n_1)$ holds,

$$TC_1^N = p\bar{x}_1^N(p) + \int_0^p \rho dF_1^N(\rho) \tag{9.51}$$

$$TC_1^D = pn_1 + \int_0^{\bar{P}_1^D(n_1)} \rho dF_1^D(\rho) \tag{9.52}$$

Hence,

$$\frac{\partial TC_1^N}{\partial p} = \bar{x}_1^N(p), \quad \frac{\partial TC_1^D}{\partial p} = n_1 \tag{9.53}$$

$$\frac{\partial TC_1^N}{\partial n_1} = 0, \quad \frac{\partial TC_1^D}{\partial n_1} = p - \bar{P}_1^D(n_1) \tag{9.54}$$

Total differentiation of objective function is

$$\frac{\partial TC_1}{\partial p} = (1-\gamma)\bar{x}_1^N(p) + \gamma n_1 \tag{9.55}$$

$$\frac{\partial TC_1}{\partial n_1} = -\gamma\left(\bar{P}_1^D(n_1) - p\right), \quad \frac{\partial TC_1}{\partial n_2} = 0 \tag{9.56}$$

$$\frac{\partial TC_2}{\partial p} = n_2 \tag{9.57}$$

$$\frac{\partial TC_2}{\partial n_1} = 0, \frac{\partial TC_2}{\partial n_2} = -(\bar{P}_2(n_2) - 1) \tag{9.58}$$

The Lagrangian is defined as

$$\mathcal{L} = TC_1 + TC_2 - \lambda(p - M) \tag{9.59}$$

The first-order condition is

$$\frac{\partial \mathcal{L}}{\partial p} = \frac{\partial TC_1}{\partial p} + \frac{\partial TC_2}{\partial p} - \lambda\left(1 - \frac{\partial M}{\partial p}\right) = 0 \tag{9.60}$$

$$\begin{aligned}\frac{\partial \mathcal{L}}{\partial n_1} &= \frac{\partial TC_1}{\partial n_1} + \frac{\partial TC_2}{\partial n_1} + \lambda \frac{\partial M}{\partial n_1} \\ &= \frac{\partial TC_1}{\partial n_1} + \lambda \frac{\partial M}{\partial n_1} = 0\end{aligned} \tag{9.61}$$

As

$$\frac{\partial \mathcal{L}}{\partial n_2} = \frac{\partial TC_1}{\partial n_2} + \frac{\partial TC_2}{\partial n_2} + \lambda \frac{\partial M}{\partial n_2} = \frac{\partial TC_2}{\partial n_2} + \lambda \frac{\partial M}{\partial n_2} \leq 0 \tag{9.62}$$

is always satisfied, $p = \bar{P}_2(n_2)$ must hold.

Now consider the case that $p = 1$, $\bar{P}_1^N(n_1) = 1$ and $\bar{P}_2(n_2) = 1$, i.e. $x_1^N(p, n_1) = n_1 = 1 - a_1^N$, $x_2(p, n_2) = n_2 = 1 - a_2$ and $m_1 = 1$ holds. In that case,

$$\frac{\partial \mathcal{L}}{\partial p} = (1 - \delta a) + (1 - a) - \lambda\left(1 - \frac{\mu(1-\gamma)\delta a}{(1+k)(1-\delta a)^2}\right) \tag{9.63}$$

$$\frac{\partial \mathcal{L}}{\partial n_1} = -\gamma(\delta - 1) + \frac{\lambda\mu(1-\gamma)}{(1+k)(1-\delta a)} \tag{9.64}$$

When (9.37) holds, for λ which satisfies $\partial \mathcal{L}/\partial p = 0$, $\partial \mathcal{L}/\partial n_1 < 0$. This means that the increase in the supplying capacity and the increase in price decrease the total cost of water acquisition. Therefore, when the first-order condition (9.60) – (9.62) is satisfied, $p^{**} > 1$, and $\bar{P}_1^N(n_1) < p^{**} \leq \bar{P}_1^D(n_1)$ and $\bar{P}_2(n_2) = p^{**}$.

(Proof of Proposition 2)

The logic of the proof is basically the same as that for **Lemma 2**. Again, $p > \bar{P}_1^D(n_1)$ and $p > \bar{P}_2(n_2)$ cannot be the optimal. When $p \leq \bar{P}_1^N(n_1)$ and $p \leq \bar{P}_2(n_2)$, the zero profit constraint requires . Then, is optimal. But as in **Proposition 1**, any $n_2 \in [0, 1 - a_2]$ is optimal.

Thirdly, a point (p, n_1, n_2) that satisfies (9.36) and $\bar{P}_1^N(n_1) \leq p \leq \bar{P}_1^D(n_1)$ and $p \leq P_2(n_2)$, the Lagrangian, is written as

$$\mathcal{L} = TC_1 - \lambda(p - M). \tag{9.65}$$

Then the first-order condition is

$$\frac{\partial \mathcal{L}}{\partial p} = \frac{\partial TC_1}{\partial p} - \lambda\left(1 - \frac{\partial M}{\partial p}\right) = 0 \tag{9.66}$$

$$\frac{\partial \mathcal{L}}{\partial n_1} = \frac{\partial TC_1}{\partial n_1} + \lambda\frac{\partial M}{\partial n_1} = 0 \tag{9.67}$$

As

$$\frac{\partial \mathcal{L}}{\partial n_2} = \frac{\partial TC_1}{\partial n_2} + \lambda\frac{\partial M}{\partial n_2} < 0 \tag{9.68}$$

is always satisfied for any $\lambda > 0$, $p = \bar{P}_2(n_2)$ holds.

Comparing (9.60) and (9.66), the difference in the first-order condition between the case of partial authority allocation and that of impartial authority allocation is the exclusion of the effect of the price change on the total cost of water acquisition in Region 2, i.e. $\partial TC_2 / \partial p$. As $\partial TC_2 / \partial p > 0$ is holds, the optimal price $p^{\circ\circ}$ is always larger than the optimal price under the case of impartial allocation p^{**}.

10 Asymmetric ownership between municipalities in water supply system provision in Indonesia

Mangapul L. Nababan[1], *Masamitsu Onishi*[2] *and Kiyoshi Kobayashi*[3]

[1] Center for Strategic Studies, Ministry of Public Works Indonesia (Jalan Pattimura 20 Kebayoran Baru Jakarta Selatan 12110, Indonesia) E-mail: mangapul_nababan@pu.go.id

[2] Graduate School of Engineering, Kyoto University (C1-333, Kyoto University Katsura Campus, Nishikyo-ku, Kyoto 615-8540, Japan)E-mail:onishi.masatmisu.7e@kyoto-u.ac.jp

[3] Graduate School of Management, Kyoto University (Yoshida Honmachi, Sakyo-ku, Kyoto 606-8501 Japan) E-mail: kobayashi.kiyoshi.6n@kyoto-u.ac.jp

Abstract

Following the decentralization process in Indonesia, the role of local governments has become more important in terms of developing policies for regional public infrastructure. Water supply system provision has distinguishing characteristics in terms of jurisdictions and regions, which affect the quality and quantity of water delivered to residents. Management of PDAM-Local Enterprise in water supply system provision is complex, especially when the formation of new municipalities results in a need to separate assets in a fair way. In the case of PDAM Tirta Raharja, ownership of PDAM belongs to Bandung Regency, whereas its service coverage includes Cimahi City and West Bandung Regency. Such asymmetric ownership has been investigated in order to understand its implications for the decision making process, particularly in matters related to expanding service coverage and improving the efficiency of public facility provision.

Keywords: Asymmetric ownership, municipalities' cooperation, water supply system

10.1 INTRODUCTION

One global agenda among Millenium Development Goals (MDGs) is to halve, by 2015, the proportion of people without sustainable access to safe drinking water and basic sanitation (Target 7c). The government of Indonesia has boosted this MDG target, aiming to achieve 75.29% access in urban areas and 65.81% in rural areas (Bappenas, 2010). Current conditions, at 49.82% and 45.72% in urban and rural areas, respectively, have prompted the government of Indonesia to mark these targets with a critical note: "need special attention" (Bappenas, 2010). In light of these conditions, the Indonesian government has also highlighted that in general, access to the water supply system, usually provided by PDAM, tends to be higher for households in urban than in rural areas. Data regarding current achievement shows that the slow pace of improvement in the water supply system is due to low development as well as low investment in this sector, especially in cities. Other issues are the poor maintenance and management of institutions responsible for water supply system provision.

To provide water and sanitation according to MDG goals, it has been estimated that $30 billion a year is required, a number criticized in 2006 in *The Economist,* which raised the question of who should be responsible for paying the provision cost and who should administer the service. Under the current conditions of democratization and decentralization in Indonesia, however, it is obvious where the answer can be found: local governments.

With the recent transformation of many Asian countries' political paradigms, including Indonesia's, into decentralized systems, the role of local governments in promoting access to safe and clean water with improved piped water has become very important, particularly in urban areas. However, the transition process from centralization to decentralization has caused issues to emerge about the capacity of local governments to provide local public goods. The lack of local budget resources, adaptation to a new bureaucracy system as well as a new political process at the local level have somehow influenced urban infrastructure improvement.

For many local governments, PPP might be one option for water supply provision. However, such issues as feasibility and a lack of water sources due to geographical conditions have been obstacles to attracting new investors.

Our main focus in this research is to identify problems of water supply system provision associated with local governments in the decentralization era. Particularly, we investigate how neighboring local governments provide water supply systems where asymmetric ownership exists. Asymmetric ownership here can be defined as one of the results of the decentralization process where new autonomous regions have been formed and certain regional assets, such local enterprises, remain under inter-jurisdictional management. We will take as an example PDAM Tirta Raharja (a local enterprise in water supply system provision), in Bandung Regency and Cimahi City, West Java Province, Indonesia. We aim to formulate an appropriate statement of problems facing two local governments, and to propose solution concepts as remedies.

Our paper is structured as follows. Section 2 describes water supply system policy in Indonesia after decentralization. Section 3 gives a brief description of our study area in Bandung Regency and Cimahi City. Section 4 analyzes the issues according to the information and data collected, and synthesizes clarification based on results of an interview with the manager of PDAM Tirta Raharja, Bandung Regency, a policy maker in Cimahi City. Section 5 formulates a problem statement and expected solutions. Section 6 concludes the paper.

10.2 SOCIO-ECONOMIC CONTEXT OF WATER SUPPLY IN INDONESIA

10.2.1 Decentralization Policy in Indonesia

Water supply system provision is a part of water management that requires careful treatment from society. It is conventionally considered that the water supply system should be managed by the public sector, as it is accompanied by a large initial investment, which under private ownership might increase the possibility of a natural monopoly. Therefore, the water supply system is inevitably managed under the national administration system in each country.

The decentralization policy in Indonesia promoted around the end of 20th century had a great impact on the management of the water supply system. According to Bird and Vaillancourt (1998), decentralization applied to the governance system of Indonesia can be defined as *the transfer of significant degrees of authority and responsibility for governmental expenditures and revenues from central government to lower levels of government*. This new system is considered to represent a significant change in the Indonesian governance system.

With the enactment of Law No. 22/1999 on Regional Administration, the national administration system transferred more authority to regional governments, called *kota* (cities) and *kabupaten* (regencies). By implication, city mayors (*walikota*) and regents (*bupati*) do not need to report to the governor of the province they belong to, as under decentralization they should be responsible to their residents represented in the house of representatives (DPRD). A change in provincial level also implies that the governor is currently portrayed as a representative of the central government in his/her province, with his/her main tasks being coordination with other municipalities. A province also has its own house of representatives.

Since the 1950s, Indonesia has been globally known as one of the biggest centralized countries. The previous regime justified the centralization policy as the sole way to maintain the unity of the nation, which is very rich in islands, population, ethnicities and cultures. The centralization policy at that time was considered repressive and neglectful of local political aspiration (mayors, regents and governors were elected by the central government), resulting in regional disparities and favoritism issues in terms of regional development. Local communities have long insisted that they deserve more authority and have demanded to be given more opportunities in terms of participation in development on a local scale. There have been discussions of the concepts of the decentralization policy since the 1970s (Bahl *et al.* 2001) at the national level, followed by the establishment of pilot programs where some local governments were given more responsibilities compared to others. Soon after Reformasi 1998, the decentralization policy was enacted as part of the governance reform agenda.

Law No. 22/1999 on Regional Government (improved with Law No. 32/2004) has changed the hierarchical relationship among central governments, provincial governments and city/regency governments. Taking infrastructure development, for example, in the centralization period, all budget proposals for local infrastructure development were subject to approval by the central government. After decentralization, as shown in Figure 10.1, it became possible for the regional budget for infrastructure development to come from several sources, both from the fiscal decentralization mechanism and regional revenues. Consequently, provincial governments maintained a direct line with the central government while local governments enjoyed more independence. This change was also followed by the transfer of various autonomous authority from the central government to local government, including, for instance, infrastructure management. Figure 10.1 below depicts the relationships associated with the infrastructure budgeting process in the decentralization era. This system has newly introduced a relatively limited role for provincial government, in the inter-municipal coordinated management of, for example, rivers and roads. The system led to greater stability, as due to Law 32/2004, province governors, regency regents and city mayors came to be elected by their constituents in regional direct election processes.

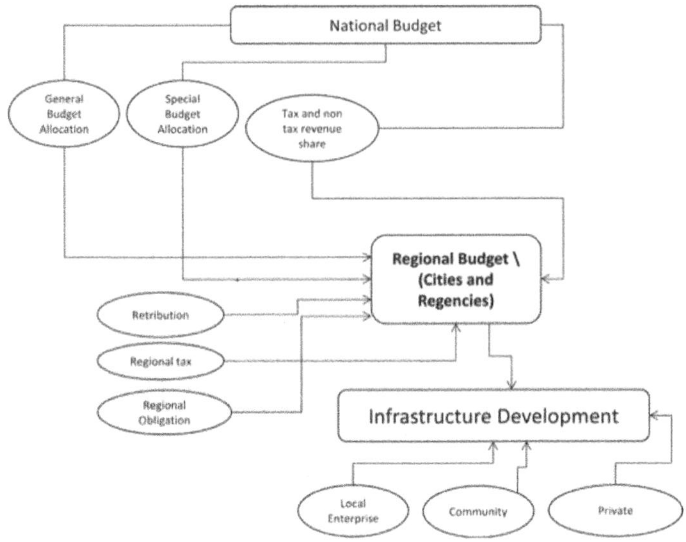

Infrastructure budgeting in decentralization era based on Laws on Regional Administration.

10.2.2 Water Resources in Indonesia

Change in the governance system has naturally led to the reform of certain strategic regulations related to infrastructure management. Indonesia's Water Resources Law (Law No.7/2004) defines water for basic needs as a basic water right, by establishing the state's responsibility to guarantee water for rudimentary needs (ADB, 2008). Article 5 states that *"the State guarantees the right for every person in obtaining water for minimum rudimentary daily use to fulfill a healthy, clean and productive life"*.

The Law also emphasizes that the right to access water for minimum daily basic needs is guaranteed by the state (Al'Afghani, 2006). The state under the Law has the obligation to provide water for everyone under Indonesian jurisdiction, by delegating the duty to city/regency governments.

Water uses are distinguished either as communal water use rights or commercial water use rights. Communal water use rights are mainly given for basic needs and noncommercial purposes, including domestic/household uses and traditional agriculture, which are excluded from formal permit requirements. Due to the climatic characteristics of Indonesia, in times of shortage, especially in the dry season, water use is prioritized for domestic use and agriculture in existing small-scale irrigation systems. Commercial use rights are considered exploitation uses for mainly commercial purposes that obviously need formal licensing. The Law also guarantees 'water exploitation rights,' which can be given to individuals or enterprises with a permit from the central or regional government.

Another important term in the law is *privatization,* which is defined as the selling of the government's shares of a state-owned enterprise to other parties to accelerate or boost the performance and value of the company and to expand share ownership for the public (Al' Afghani, 2006). Water use for PDAMs can be categorized as commercial uses at the local level, with authority being given by a particular local government to use water for clean water

Water rights and licensing.

Type of rights	Method to obtain	Target	Purpose
Communal water use rights	Without license < 2 litres/s	Individual	Basic Daily needs
i.e. basic needs and non-commercial		Individual or groups	Community, Agriculture in existing irrigation system
	With license	Individual	Basic daily needs
		Individual or group	Agriculture outside existing irrigation system
		Group	Basic Daily needs
Commercial water use rights	With license	Individual/Group/ Organization	Need for business or commercial use

Source: Law No.7/2004, Dir. Water Resources and Irrigation, Bappenas.

production as well as for distribution. Water use for households and industries is regulated in the Law, which can be seen in Table 10.1 above.

Due to a lack of budget resources for infrastructure investment and the need to improve good governance, a local government can arrange cooperation with a third party, especially for water provision that is usually a monopoly (Decree of Minister of Internal Affairs 43/2000). According to Government Regulation No. 14/1987, water supply systems, particularly in urban areas, are provided by PDAMs, in semi-autonomous joint management where the local government owns shares partially.

By November 2010 (Perpamsi, 2012), there were 402 entities of water supply system providers in Indonesia. The three categories of the entities consist of the following: i) local enterprises (PDAMs); ii) water supply boards/units under local governments; and iii) private providers. PDAMs are distributed to 83 cities and 319 regencies, in which they can be categorized according to the number of customers they have, as follows: 12 PDAMs with more than 100,000 customers, 19 PDAM with 50,000 – 100,000 customers, 20 PDAMs with 30,001 – 50,000 customers, 136 with 10,001 – 30,000 customers, and 212 with less than 10,000 customers. However, based on a performance evaluation by BPP SPAM-Supporting Board of Water Supply System Provision (2010), there are 103 PDAMs with 'healthy' performance, 115 PDAMs marked as 'not healthy' and 119 PDAMs classified as 'ill'. BPPSPAM (2010) also stated that there are 8,032,099 customers who accessed no PDAM water, representing 24% of national coverage service distributed in urban areas (47%) and rural areas (11%).

10.2.3 Water Supply System Administration in Indonesia

Local government has the authority to make policies on PDAM water service level as well as price. In addition, PDAM plays a role in business operation by determining raw water availability, raw water transmission, water treatment facilities, clean water transmission, reservoir unit development to water distribution to customers. Briefly, the local government has ownership of and rights over water resources, where some of the authority to determine the raw water amount needed for production is given to PDAM. However, there are many regulations, most involving different sectors and government layers, that should be taken into account.

In general, water supply system administration applied in all cities and regencies of Indonesia is similar and can be classified according to government level. At the central level, PDAMs are expected to deal with ministries responsible for health issues (Ministry of Health, Ministry of Environment), water resources and drinking water issues (Ministry of Public Works), and planning and investment issues (National Development Planning Board and BPPSPAM). At the provincial level, due to regional sustainability issues, PDAMs need to coordinate with provincial offices associated with water resources, mining and forestry. Last, at the local level, PDAMs operationally are required to coordinate with the local planning division, environment division, city/regency revenue division and health division. In addition, such issues as service capacity and price determination also require approval from the House of Representatives.

As summarized by ADB (2008), local governments have faced several issues related to water supply system provision since the stipulation of the Law of Water resources and Decentralization. We will explain in more detail, as follows.

a. Groundwater licensing is operated by district (municipalities). PDAM water provision through the use of surface water requires a permit from the provincial government. However, ambiguity remains because in practice, in some regions, provinces operate the permit system under previous law while in others permits are given under current provincial regulations.
b. The law does not permit water trading. However, water commercialization is possible through the establishment of a state-owned company as operator. For example, some state-owned companies are given authority over river water and can provide water to PDAMs.
c. The decentralization process has led to a lack of coordination between stakeholders, for example in the spatial planning and water resources planning processes. The formation of new autonomous regions raises issues of inter-jurisdictional coordination in the water supply system. This is very complicated due to the different points of view of local leaders and the many political interests involved.
d. Insufficient cost recovery from commercial users. In the case of PDAMs, not every customer can afford the price of water; consequently, PDAMs apply block prices where non-commercial users are subsidized. However, many PDAMs face problems with cost recovery nowadays, due to increasing production costs and the difficulty of raising prices. Some local governments provide subsidies, but budget allocations are insufficient to cover the cost. Previously, the central government could intervene, but since decentralization, local governments have been responsible for overcoming these kinds of issues.

10.2.4 Challenges

In the case of water supply provision, the authority to determine the production level and price is based on the regulation of the regent or mayor of the city. The Decentralization Law dictates that the authority to exploit water is determined according to the area of jurisdiction; however, due to characteristics of geographical areas, some regions still rely on other regions to supply raw water as input for clean water production. In this sense, local government has the authority to use the water resource, for example the surface water, groundwater, and spring water in its jurisdiction, to supply clean water for its residents. For surface water such as that of a river involving two or more regions, the river can be considered as a provincial river under provincial government authority, but for other sources such as groundwater and spring water, the permit system is more complicated due to inter-sectoral issues, which necessitate coordination among forestry and mining offices at each respective government level. Hence, water use rights regulation in terms of cross-jurisdiction water transfer is still under debate, and until now, conflict between jurisdictions has been resolved solely through bilateral or multilateral agreement. However, in the case of conflict between cities or regencies, the provincial government can play the role of mediator and facilitator, as stipulated in the law.

An effective political process is needed to meet the targets of local government, which plays the role of providing water for its residents. Local government usually establishes local enterprises such as PDAM to fill such obligations. PDAM itself is a semi-private company, partly owned by the local government but with private-oriented objectives. PDAM mainly has a monopoly privilege of providing piped water and distributing it with household connections. However, determination of the production level and water price is still the domain of the local government, which depends on the political process to achieve its goals.

The central government has initiated certain actions to improve water supply system management, for example by facilitating for local governments the process of attracting investors to potential PDAMs. Regarding asset sharing as part of new region formation, several actions have been undertaken to control potential conflict over water and to improve the quality of coordination and cooperation between stakeholders. Nevertheless, disputes between local governments concerning asset sharing and asset management still occur occasionally. In the next section, we will discuss a case in the Bandung region where such issues have emerged.

10.3 A CASE STUDY OF PDAM TIRTA RAHARJA (BANDUNG REGENCY AND CIMAHI CITY)

This section gives a brief description of Bandung Regency and Cimahi City, with particular focus on how residents in both regions access clean water. This section also provides a profile of PDAM Tirta Raharja.

10.3.1 Description of Bandung Regency and Cimahi City

As of 2006, Bandung Regency and Cimahi City had populations of 4,399,788 and 520,035, respectively. Bandung Regency constitutes one of the biggest regencies in West Java, 327.434 Ha with a density of about 1,391 people/km^2. Historically, Cimahi City and Bandung Regency comprised a single unit called Bandung Regency. Until 2001, people normally thought of Bandung City, the capital of West Bandung Province, and Sumedang Regency as parts of Bandung Greater Area, or BMA. BMA itself consists of Bandung Metropolitan City and the urban areas surrounding it.

After decentralization, one emerging issue on inter-local government provision of water supply system involved Bandung Regency and Cimahi City. Cimahi City, a new autonomous region formerly united with Bandung Regency, claimed that it had contributed largely to the profit of PDAM Tirta Raharja due to the fact that a significant portion of the total number of customers were registered residents of Cimahi City. In order to give brief information about the location of our study area, we include the following map, Figure 10.2, which shows the location of Bandung Regency and Cimahi City in the context of Indonesia.

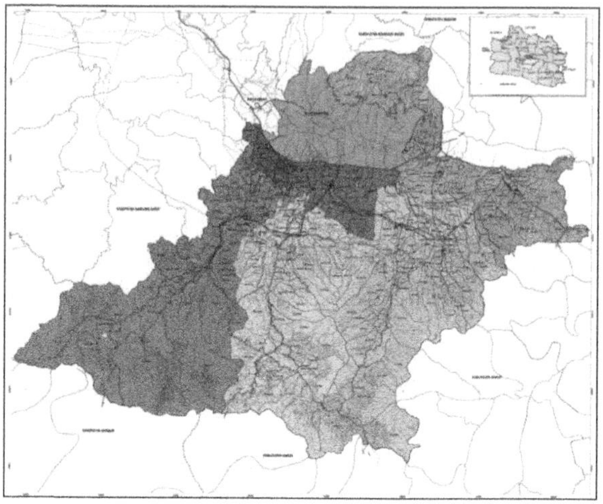

Bandung Regency includes all colored regions except for those colored green (Bandung City).Cimahi City is the white region to the left of Bandung City. (*Source*: Bandung Regency Regional Spatial Plan, 2006).

PDAM Tirta Raharja supplies water to Bandung Regency, West Bandung Regency and Cimahi City. Due to the lack of access to PDAM because of limited connections, PDAM water prices, or the poor quality of private wells, some regions in Bandung Regency have found it necessary to seek alternative water sources. In order to provide adequate water supply, some communities have proposed cooperative actions with local governments to realize a community-based water supply system. Generally, local governments provide budgets for developing physical infrastructure for water supply systems, while operation and management are left to communities. By 2005, 219 of 440 villages in Bandung Regency had established community-based water supply systems.

Also important to consider in developing water supply systems are issues related to water sources, which vary according to regional characteristics in Bandung Regency. Usually, communities attempting to establish community-based water supply systems should find the necessary adequate water source. Some communities rely on spring water, artesian wells, or even PDAM water if available, whereas the construction of reservoirs or transmission and distribution systems could be provided by local governments using particular assets such as available land or private land rented through a profit-sharing system. Figure 10.3 shows access to drinking water by households in Bandung regency.

Located near Bandung Metropolitan City, Cimahi City has been considered a satellite of Bandung for decades. The city has grown rapidly, especially since 2001, when Cimahi City legally became autonomous, thus requiring more investment in infrastructure, particularly for water supply.

Based on a population survey in 2000, the population of Cimahi City was 541,139, with a 2.06% per year growth rate. With a population density of 13.134 person/km^2, the average

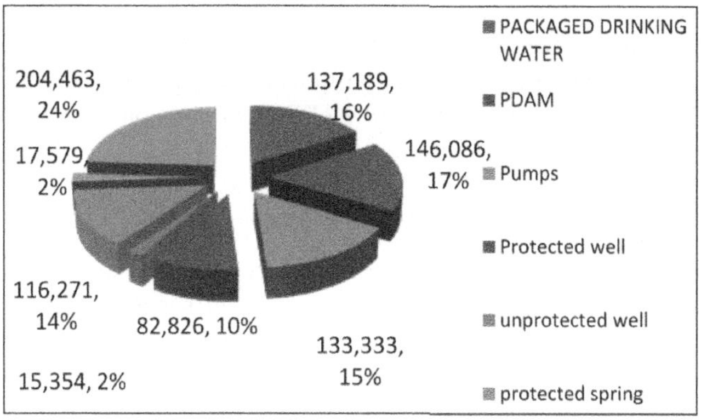

Access to drinking water by households in Bandung Regency. (*Source*: PDAM Tirta Raharja Figures 2006, PDAM Tirta Raharja, 2006).

family numbered 3.63/household. PDAM Tirta Raharja supplies clean water to 13 of 15 districts in Cimahi City, with 14,144 connections for approximately 95.246 persons, or about 17.6% of the total population. As for categories of consumers, households accounted for 94% of total registered customers, followed by businesses at 3.48%, military bases at 1.10%, with the remainder being for miscellaneous use, such as public hydrants, social institutions, government facilities, industry, and others.

A study conducted by EHRA in 2011 (Cimahi City, 2011) shows how residents access clean water in Cimahi City, as shown in Figure 10.4 below.

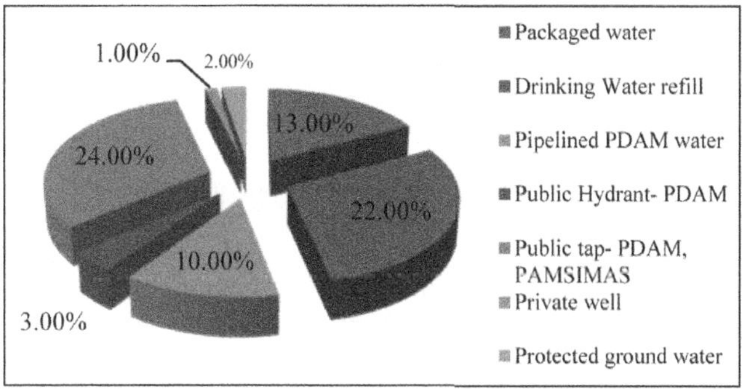

Access to drinking water by households in Cimahi City. (*Source*: Sanitation White Book of Cimahi City, Cimahi City Local Government, (2011)).

A study by Tukimin (2004) conducted in Cimahi Selatan showed that the quality of PDAM water consumed by households is good. The indicator of water quality itself is based on physical traits, that is, whether the water is colorless, odorless, and tasteless. In terms of quantity, due to the limited capacity of water provided by PDAM, several alternatives are used, among them private wells, shared connections, public hydrants, water vendors and neighbor connections. It is very common for households to have more than one water source, due to the limited capacity of PDAM and the sporadic continuity of service, for example when PDAM cannot provide 24 hours of service. Particularly in the dry season, the continuity of

PDAM service decreases, with periodic services applied in durations of 2 days, 12 hours, and 6 hours. Such conditions force customers to find alternative sources of water. In his study, Tukimin (2004) stated that 60% of PDAM consumers in Cimahi Selatan also have access to alternative water sources, whereas among 30% of PDAM non-consumers, some households have more than 2 alternative water sources.

10.3.2 History of PDAM Establishment

In the decentralized era, the role of PDAMs as local enterprises is to provide service for the residents of the city/regency. PDAMs are pushed to earn sufficient revenue to cover costs and expenses incurred, such as payment for operational and maintenance expenses, interest cost and principal past due capital goods expenditure (including rehabilitation cost), and to make contributions to Regional Income (PAD). Usually operational revenue originates from water revenues that consist of water tariff revenue and fixed charges, and non-water revenues that consist of new connection fees, penalties and others. However, the largest portions of total operational revenues come from water tariffs and fixed charges.

PDAM Tirta Raharja has been in existence since 1926, and was previously named Water Leiding Bedriff (in Dutch). Dating back to the colonial era, it was initially proposed to provide clean water for the Dutch community in Cimahi and Lembang. Then in 1977, based on Regional Policy No. XVII, followed by the Governor of West Java Decree No. 510/H.K/011/SK/77, the water company was changed into PDAM Kabupaten Bandung (Bandung Regency) with objectives both to provide clean water for Bandung Regency residents and to contribute to regional income. According to these regulations, ownership of PDAM Tirta Raharja is held by the local governments of Bandung Regency. The regulations also stated that the regent has the authority to select the PDAM Directors Board and to establish supervision boards, whose members are also selected by the regent.

PDAM Tirta Raharja has coverage service for Bandung Regency and Cimahi City with support from 6 branches which provide one region or more, according to distance (see Table 10.2 below).

Number of connections and people serviced by PDAM Tirta Raharja.

Branch	Amount of Connections	Number people serviced
Soreang	10,887	76,552
Ciparay	11,306	80,834
Rancaekek	5,126	31,660
Lembang	4,599	32,910
Padalarang	3,208	24,930
Cimahi	14,144	95,246
TOTAL	49,270	342,132

Source: PDAM Tirta Raharja Figures 2006, PDAM Tirta Raharja, 2006.

Meanwhile, Table 10.3 shows water sources used by PDAM Tirta Raharja as output for producing clean water.

Water source available and water used for PDAM Tirta Raharja.

Area (Water Source)	Available Capacity (l/s)	Used Capacity (l/s)
Soreang (River)	60	55
Banjaran (River)	40	26.58
Ciwidey (Spring)	20	0.78
Pangalengan (Spring)	30	25.58
Ciparay (River)	200	167.32
Pacet (Spring)	5	2.29
Paseh (Spring)	5	0.3
Cicalengka (Drilling Well)	8	3.42
Rancaekek (Drilling Well)	24	19
Cisarua (River)	36	22.65
Lembang (Spring)	38	24.33
Batujajar (Spring)	30	20
CikalongWetan (Spring)	8	6
Cililin (River)	20	10
Padalarang (Drilling Well)	28	20.5
Cimahi (River)	197	147.5
TOTAL	749	551.25

Source: PDAM Tirta Raharja Figures 2006, PDAM Tirta Raharja, 2006.

Usually a water company can review its prices periodically, for instance every three or five years, to adjust its profit. However, considering the ownership of PDAM, it is very difficult to increase tariffs every year to keep up with the rate of inflation. Due to such a condition, the sustainability of PDAM should be supported by the financial systems of local governments. In addition, PDAM also suffers from so-called non-revenue water issues, 35% on average (BPPSPAM, 2010). Further affecting profit conditions, PDAM Tirta Raharja was also required to contribute 55% of its profits to Bandung Regency Local government as a contribution to regional income (PAD). Figure 10.5 shows profit/loss of PDAM Tirta Raharja (1996–2003).

In 2009 PDAM Tirta Raharja gradually increased the tariff to Rp 2,100/m3, applicable to customers in the entire service area, including in Bandung Regency itself, Cimahi City and West Bandung Regency. This decision was based on Regent of Bandung Regulation No. 52 2008. However, as a consequence of the tariff increase, the agreement on profit sharing with the other two regions is still in the negotiation process. Table 10.4 and 10.5 illustrate tariff issues of PDAM in Indonesia.

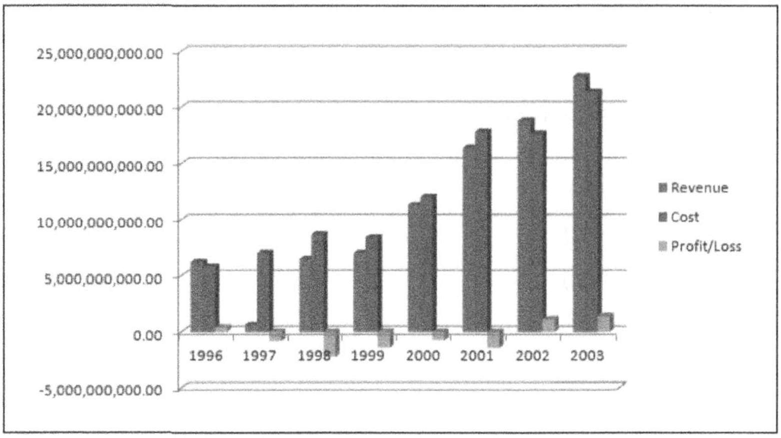

Profit/Loss of PDAM Bandung Regency 1996–2003. (*Source*: PDAM Kota Bandung Tariff Adjustment Proposal, USAID, 2006).

Water price comparison of several water sources.

Clean Water Sources	Price
PDAM	Rp 3/l
Cart Water	Rp 50/l
Gallon Water	Rp 500/l
Public Hydrant	Rp 25,000/month

Source: PDAM Kota Bandung Tariff Adjustment Proposal, USAID, 2006.

Water price comparison of several PDAMs in Indonesian Cities (Rp/m^3).

Name of PDAM	Low	Average
Bandung regency	1,000	3,500
Bandung City	560	2,014
Jakarta TPJ	4,367	6,000
Makassar City	1,000	4,500
Semarang City	1,608	4,500
Sidoarjo Regency	1,350	3,700
Tangerang regency	1,850	4,500

Source: PDAM Kota Bandung Tariff Adjustment Proposal, USAID, 2006.

10.4 GOVERNANCE OF PDAM-BANDUNG REGENCY AND CIMAHI CITY CASE

10.4.1 Governance of PDAM Tirta Raharja

We will explore a case of water supply system provision in Bandung Regency and Cimahi City, two neighboring municipalities in West Java Province, Indonesia. Both local governments are still in dispute as to how to manage the assets of PDAM Tirta Raharja, because since 2001 Cimahi City has been separated from Bandung Regency due to the process of decentralization, which implies the need for separation in the management of assets, including the water company.

As shown in Figure 10.6, the decision making process of PDAM Tirta Raharja involves only the regent of Bandung Regency. In this sense, there is no room for Cimahi City government to be involved in strategic decision making, for example in determining output service level or tariff.

Some characteristics of these regions are explained as follows. Bandung Regency and Cimahi City are different in terms of water source availability. Compared to Cimahi City, Bandung Regency has many alternatives in terms of how its residents can obtain clean water other than PDAM. If PDAM wants to make more money, it will expand by investing in Cimahi City because Cimahi City has more potential consumers. Table 10.2 and Table 10.3 show that Cimahi City's contributions, both as a water source supply to PDAM Tirta Raharja and in terms of the considerable number of customers who obtain water from PDAM Tirta Raharja, are significant. The fact that around 28.7% of the connection belongs to Cimahi City residents proves the importance of Cimahi City as a source of revenue for PDAM Tirta Raharja.

People in dry seasons suffer shortages in water supply, but Cimahi City still has an adequate and stable water supply system. However, Cimahi City cannot control the PDAM water supply, even though people suffer low quality of water from private wells, community-based, springs, etc. They want to invest in water supply systems to avoid the effects of the dry season, but even though Cimahi city government wants to expand capacity, they are unable to take action due to the lack of authority caused by the PDAM ownership issue.

In order to increase capacity as well as coverage services, local governments in Cimahi City have repeatedly proposed to Bandung Regency that the water supply to Cimahi City

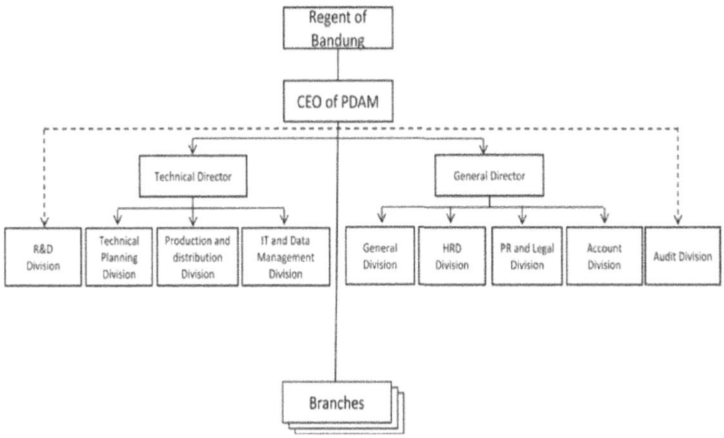

Organization of PDAM Tirta Raharja.

be increased. However, the Bandung Regency local government declined these proposals and requests due to undisclosed reasons. In addition to that, another critical issue on PDAM asset management has also been emerged. As a consequence of formation of Cimahi City autonomous region, part of PDAM Tirta Raharja which initially belong to Bandung Regency before decentralized) and are located in Cimahi City, should be transfered to Cimahi City new local government. We noted that these assets include the main office of PDAM Tirta Raharja, water treatment facilities and so on. There has been a lack of negotiation at the political level where Cimahi City demanded for asset sharing in which Bandung Regency seems to be reluctant to do so. As a matter of fact, we found that the PDAM Tirta Raharja ownership dispute is ongoing, even up to the writing of this paper.

10.4.2 Issues on Asymmetric Ownership

In this paper we attempt to formulate problem statements related to ownership issues. Accordingly, we first investigate some issues to be confirmed later. Second, we report the results of interviews with key persons responsible for making decisions regarding the problems. Finally, we list our conclusions.

In the process, we identify certain issues as follows:

a. There were historical contingencies of decentralization in Cimahi City and Bandung Regency, which imply difficulties in negotiating asset sharing.
b. The real problem of the water supply system in Bandung Regency is associated with the water source problem.
c. There are issues regarding the asymmetric ownership of PDAM Tirta Raharja, which is controlled by Bandung Regency, and over which Cimahi City does not have not authority.
d. It is necessary to identify the structure of the problem to develop a theoretical model for improving policy.

In order to clarify these issues, we conducted several interviews with key persons in the Bandung region. The interviews were conducted in the middle of March 2012 while the authors were visiting Indonesia. We arranged several discussions with key persons associated with water provision research, particularly PDAM and its relationship with decentralization policy. We visited the PDAM Tirta Raharja's key person and the representative of Cimahi City government in charge of city infrastructure planning, including water supply system provision.

The main objectives of the interviews are to clarify the problem statements we formulated and to use the interview results as a basis for developing a model to describe such conditions from the point of view of economics. Further, we present implications for social welfare and economic efficiency in both municipalities with the aim of finally resolving problems from the point of view of economics.

10.4.2.1 Interview with a keyperson of Cimahi city

The results of interviews with Cimahi City officers are summarized below:

a. In attempts to resolve the need to improve water supply system provision in Cimahi City, the Cimahi City government has tried to pursue strategies to expand the capacity of PDAM Tirta Raharja in Branch Cimahi. However, officials were hindered by the lack of authority caused by the PDAM ownership issue.

b. Regarding the policy making process of PDAM, it is difficult to establish a mutual agreement between Cimahi City and bandung regency because of conflict of interest among stakeholders in both regions.
c. Regarding financial issues, Cimahi City is highly dependent on investment from the central or provincial government due to the huge initial investment needed. Another difficulty is the assessment of the central government about the dubious feasibility of PPP implementation for the establishment of a new independent PDAM for Cimahi City, due to lack of water sources.
d. The provincial government has been involved in the plan for an independent PDAM for Cimahi City, which is also supported in the provincial framework, together with the concept of Clean Water Provision System. However, as a temporary solution, the provincial government supported Cimahi City in promoting the community-based water provision that already exists.
e. Recently, PDAM Cimahi City service level has decreased due to rapid population growth. It is apparent that non-consumers of PDAM rely on other water sources such as deep wells and community-based water supply system provision, though it is necessary to deal with potential water shortages during the dry season.
f. Establishment of Kelompok Pengelola Air (water group management, a community-based action initiated by local community with partial support from local government) is one attempt to solve water supply problems by involving community participation.

10.4.2.2 Interview with a keyperson of PDAM Tirta Raharja

In an interview with PDAM Tirta Raharja officers, we confirmed several issues regarding the ownership of PDAM assets between Bandung Regency and Cimahi City as follows:

a. Ownership issues of PDAM Tirta Raharja have emerged between Bandung Regency and Cimahi City local governments.
b. Responding to Cimahi City's aspiration regarding the possibility of being involved in management as well as asset transfer due to the consequences of decentralization could be a difficult process and require a long time for negotiations.
c. PDAM Asset should be considered as a separated asset of the local government due to the involvement of the private sector in provision. However, regulation on such issues has not yet been made clear.
d. Around December 2011, there was a MoU facilitated by ADB discussing the possibility of a win-win solution based on a profit-sharing mechanism among Bandung Regency, Cimahi City and West Bandung Regency. However, the process of MoU is still doubtful, due to differing political interests.
e. The financial structure of PDAM Tirta Raharja is independent of the local government accounting system. However, as a local enterprise in which a portion of company shares is owned by the local government, profit sharing and revenue are regulated. Some portions of profit go to regional income whereas other portions can be used for new investment.
f. The possibility of cooperation among local governments is low; however, cooperation among local enterprises could be one promising option. As a matter of fact, cooperation between PDAM Tirta Raharja and the local enterprise of Bandung Barat Regency has been established in recent years in some elements of water supply system provision.

10.5 PROBLEM STATEMENT AND EXPECTED SOLUTIONS

The decentralization policy in Indonesia is one promising objective of the Reform Movement of 1998. However, implementation of this policy has required strong commitment at every level of government. Shah and Thompson (2004) underlined that the main purpose of decentralization is to enable local governments to provide public services and to be even more responsive to their residents. In contrast, Firman *et al.* (2008) highlighted the negative consequences of a policy that leads local governments to be more reluctant to engage in the cooperation required to provide shared public facilities, for instance.

Due to the huge initial investment needed for establishing a water distribution network and for decreasing costs to additional consumers, it is very common globally for this public good to be provided by only one particular firm. In most developing countries, this monopoly characteristic is one of the main features of local government. In a setting of PDAM Tirta Raharja involving two neighboring local governments, Bandung Regency and Cimahi City, we have shown asymmetric ownership of PDAM. Further, by gaining an understanding of decentralization in Indonesia, we determine whether or not cooperation between local governments in the provision of such goods is efficient.

According to Slack and Bird (2007), in an attempt to reduce the cost of public good provision, voluntary cooperation between municipalities or local governments could be one possibility requiring minimal government restructuring. They argue that voluntary cooperation could be a promising solution in the decentralized era, where intergovernmental transfers have been limited due to the change from centralized to decentralized governance. They also highlight some benefits of cooperation between local governments, such as the lack of permanent agglomeration and the independent institutional status common in the USA and France. In particular, cooperation between local governments is politically easy to establish and can be disbanded relatively easily as well.

It is apparent that inter-local government has also played a significant role in the performance of budget and expenditure. Surveys by the International City/County Management Association, ICMA (1997) in the US reported that inter-local cooperation has been important in three ways in local service provision, just under in-house and private. According to Feiock et al., there is a two-stage process in establishing cooperation between local governments. First, communities decide whether to consider inter-local government cooperation by carefully taking into account the gains and effectiveness of cooperation. At the second stage, communities need to establish new institutions to accommodate cooperation, which is needed to handle bargaining and collective action issues related to inter-local agreements (Feiock *et al.* 2010).

Institutional arrangement as the result of cooperation between local governments could lead to fund policy. For example it can levy taxes or collect contributions from the municipalities or it can levy user fees to pay for services. This is interesting because cooperation could be a means for local governments to increase their capacity in public good provision, to help them to achieve economies of scale, as well as to deal with externalities issues. However, Slack (2007) emphasized that accountability issues between local governments should be considered as cooperation involving the public good provision established. To avoid such problems, it is very important to ensure that local governments involved in cooperation have the same objectives.

Another important issue in establishing cooperation between local governments is bargaining among the local governments involved. Due to asymmetric conditions, it is possible for some municipalities to have little to bargain with in their negotiations with other local governments. In the case of public good provision, for instance, such a power imbalance could be due to the limited availability of natural resources or budget constraints. In order to avoid potential conflict in the future, appropriate contracts or agreements should be applied in inter-local government cooperation.

It is challenging for decentralized local governments to resolve problems of economies of scale and inter-jurisdictional provision of public services. In a fragmented system of local governments, efficiency in public good supply also needs to be considered. In the case of asymmetric ownership between the Bandung Regency and Cimahi City local governments, we proposed inter-local cooperation as an alternative for resolving the inefficiency of water supply system provision in the Bandung region. According to an Advisory Commission on Intergovernmental Relations (ACIR), inter-local cooperation can be defined as voluntary exchange between autonomous local governments for the efficient supply of public services without surrendering their autonomy. Further, Axelroad (1984) emphasized that inter-local agreement as a result of cooperation is an attempt of individual jurisdictions to pursue their self-interest without the aid of a central authority to force them to cooperate with each other.

Local governments can cooperate with each other by using both formal and informal agreements(Zimmerman, 2007; Post, 2004). By taking US local governments as example, about 61% of cities received service under formal and informal agreements in 1972 through inter-local government action. Post (2004) stated that informal agreements can be understood as handshake deals between transacting local governments where the division of responsibility is understood, even though it is never formalized (Post, 2004). In contrast, formal agreements can be pay-for-service agreements, joint service agreements or mutual service agreements.

However, as Williamson states (Williamson, 1996), inter-local service agreements as a form of cooperation are difficult to maintain because they are influenced by behavior and environment conditions. For example, in our setting of asymmetric ownership condition, one partner may behave opportunistically to appropriate the larger share of gains. Other important considerations are the political conditions that usually lead to changes in administration in local governments that sometimes result in decisions to change or to withdraw from original agreements (Shrestha, 2008).

10.6 CONCLUSION

Since decentralization, water supply system provision in Indonesia has become more complex. Challenges vary from nature and technological issues to socio-economic issues. Water resources availability tends to be inter-jurisdictional; hence local governments nowadays are facing challenges with coordination and cooperation issues. However, decentralization also leads to difficulty in achieving agreement between respective local governments involved in water supply system provision problems, due to political constraints and conflicting interests among parties involved.

In this paper, we have introduced the asymmetric ownership of PDAM Tirta Raharja between Bandung Regency and Cimahi City. We highlighted that such ownership could be one critical factor resulting in inefficiency of water supply system provision in Bandung Region. We introduce inter-local government cooperation as a promising mechanism for resolving asymmetric ownership problems between local governments. Even though agreements between local governments can be formal and informal, agreements also tend to change due to political conditions as one consequence of the decentralization policy.

10.7 ACKNOWLEDGEMENT

We thank Mr. Hery Antasari, Head of Infrastructure Planning in Bapeda Cimahi City, and Ms. Hendriati from Technical Management of PDAM Tirta Raharja for their acceptance, hospitality and kindness during interviews.

We are also very thankful to SPPAK ITB, who helped us in arranging interviews in the middle of March 2012, especially Dr. Ibnu Syabri, Dr. Miming Miharja and Dr. Sri Maryati.

10.8 REFERENCES

ADB-Asian Development Bank (2008). Water rights and water allocation, issues and challenges for the region, *Asian Development Bank, Advance Draft for 3rd General Meeting,* Mandaluyong City, The Phillippines, *ADB*.

Advisory Commission on Intergovernmental Relations (ACIR) (1985). *Intergovernmental Service Agreements for Delivering Local Public Services*: Update 1983 (A-103). Government Printing Press, Washington, DC.

Al'Afghani M. M. (2006). Constitutional Court's Review and The Future of Water Law in Indonesia. *Law Environment and Development Journal,* **2**(1), 1–18.

Axelrod R. (1984). *The Evolution of Cooperation*. Basic Books, New York.

Bahl R., Aten R.H., and Alm J. (2001). Can Indonesia decentralize successfully? plans, problems, and prospects. *International Center for Public Policy, Working Paper Series,* AYSPS, GSU.

Bandung Regency Local Government (2006). *Regional Spatial Plan,* Bandung Regeny Local Government, Bandung.

Bappenas (National Development Planning Agency) (2010). *Report of the Achievement of The Millenium of Development Goals Indonesia, 2010,* Ministry of National Development Planning.

Bird R. M. and Vaillancourt F. (1998). *Fiscal Decentralization in Developing Countries.* Cambridge University Press, Cambridge.

BPPSPAM, *Kinerja PDAM 2009* (BPPSPAM, 2010).

Cimahi City Local Government (2011). *Sanitation White Book of Cimahi City.* Cimahi City Local Government.

Decree of Minister of Internal Affairs No. 43/2000 regarding on Guidance for cooperations between local enterprise with third party.

Feiock R. C. and Scholz J. T. (eds.) (2010). *Self-organizing Federalism: Collaborative Mechanism to Mitigate Institutional Collective Action*. Cambridge University Press, New York.

Firman T. (2008). In search of a governance institution model for Jakarta Metropolitan area (JMA) under Indonesia's new decentralization policy: old problems, new challenges. *Public Administration Studies,* **21**(2–3), 143–157.

Government Regulation (1987). *On Transfer of Partial Authority on Public Works from Central Government to Local Government,* No. 14.

International City/County Management Association (ICMA) (1997). Local intergovernmental agreements: strategies for cooperation. *Management Information Service,* **29**(7), 1–11.

Law No. 22/1999 on Regional Administration.

Law No. 32/2004 regarding on Regional Administration.

Law No. 7/2004 on Water Resources.

PDAM Tirta Raharja (2006). *PDAM Tirta Raharja Figures 2006,* PDAM Tirta Raharja.

Perpamsi (2012) Peta Masalah PDAM, *Ringkasan Eksekutif (In Bahasa Indonesia),* http://perpamsi.or.id/gallery/document/PETADES2010web-1.pdf (accessed time: May 20th, 2012).

Post S. S. (2004). *Metropolitan Area Governance and Institutional Collective Action.* In: Metropolitan Governance: Conflict, Competition, and Cooperation, R. C. Feiock, (eds.), Georgetown University Press, Washington, D.C.

Shah A. and Thompson T. (2004). *Implementing Decentralized Local Governance: A Treacherous Road with Potholes, Detours and Road Closures*, World Bank.

Shrestha M. K. (2008). Decentralized governments, networks and interlocal cooperation in public goods supply. *Electronic Theses, Treatises and Dissertations*-The Florida State University, Paper 324.

Slack E. and Bird R. M. (2007). An approach to metropolitan governance and finance. *Environment and Planning C: Government and Policy*, **25**, 729–755.

Tukimin S. (2004). Potensi Pengembangan pelayanan Air Bersih PDAM Tirta Raharja Cabang I Cimahi. *Master Thesis (in Bahasa Indonesia)*, Universitas Diponegoro, Indonesia.

USAID (2006). PDAM Kota Bandung Tariff Adjustment Proposal – 2006. *Environmental Service Program*, USAID, Jakarta.

Williamson Oliver E. (1996). *The Mechanisms of Governance*. Oxford University Press, New York.

Zimmerman J. F. (1974). *Intergovernmental Service Agreements and the Transfer of Functions in Substate Regionalism and the Federal System 3*, Government Printing Office, Washington, DC.

11 Comprehensive social participatory model for water springs conservation management in Indonesia

Adipandang Yudono[1], Arief Rachmansyah[2], Sri Sudaryanti[3], Sudarto[4]

[1] Department of Urban and Regional Planning, Brawijaya University, Indonesia. E-mail: adipandang@ub.ac.id

[2] Department of Civil Engineering, Brawijaya University, Indonesia. E-mail: ariefftub@yahoo.com

[3] Department of Aquaculture Resource Management, Brawijaya University, Indonesia. E-mail: gociunibraw@gmail.com

[4] Department of Soil Science, Brawijaya University, Indonesia. E-mail: sudarto_fpub@yahoo.co.id

Abstract

The water spring is one very useful natural resource because of its use as water supply in rural and small municipal areas. In the last decade, the quality and quantity of some water springs have been declining, with constant exploitation further impeding the conservation of natural resources. Until today, water spring conservation programs have relied on a biophysical approach (landscape), without regard to social considerations. The community of water springs users has received little attention. The biophysical approach has proved ineffective in solving problems, and natural resource conservation action has shown no improvement. This paper explains the experiences of the community in Malang District, Indonesia, in their efforts to conserve the water springs that they need. One effort toward water spring conservation management is implementing a Comprehensive Social Participatory Model.

Keywords: conservation, management plan, public participatory, water spring.

11.1 INTRODUCTION

Drought disaster in the dry season, landslides upstream, and downstream flooding in the rainy season in the catchment area system are serious indications of the necessity to address water problems (Hammer and Hammer Jr, 2004). In research on natural resources, there has been too heavy a reliance on the biophysical approach (landscape), and a lack of consideration of the social aspects of natural resources. This sectoral approach has proved ineffective in solving the technical problems of natural resources, with too little attention being paid to resource capacity for human users. As a result, from year to year, circumstances surrounding natural resources have become even more alarming. Lack of available social data regarding water spring areas makes it difficult to plan their management, with problem solving being carried out not based on facts in the field. Community involvement in the decision-making process is less than maximal, so that support for the program's success is low. Implementation of the program has performed well, but maintenance is often neglected due to low community involvement. According to Law No. layout. 26 years 2007 in Chapter VII, the public has the right, obligation and role to organize its own areas, while according to article 84 of the Water Resources Act No. 7, 2004, communities have an equal opportunity to participate in the process of planning, implementation, and monitoring of water resources management. According to another regulation, Indonesia's Environmental Ministry Decree No. 32, 2009, on the conservation and management of Environment Article 70, paragraph 3, the role of the community is to increase awareness of the conservation and management of the environment, enhance self-reliance, empower the community, foster partnerships, develop capabilities, and strive for a pioneering community. To conserve and manage water springs, a study of social aspects is needed. Furthermore, Hakli and Minca (2009) examine "*locating' or 'finding' trust in social practises related to economic transactions, democratic government and policy formation*". One of the efforts of the management through social studies is implementing a Water Spring Management Participatory Model towards policy formation. The scope of this study in Malang Regency (see Figures 11.1) includes 34 water springs managed by Indonesia's Drinking Water Regional Bureau (PDAM Kab. Malang), all located in the Brantas river basin, with three locations selected as the highest priority water resources for implementation of the Water Spring Management Participatory Model (see Figure 11.2).

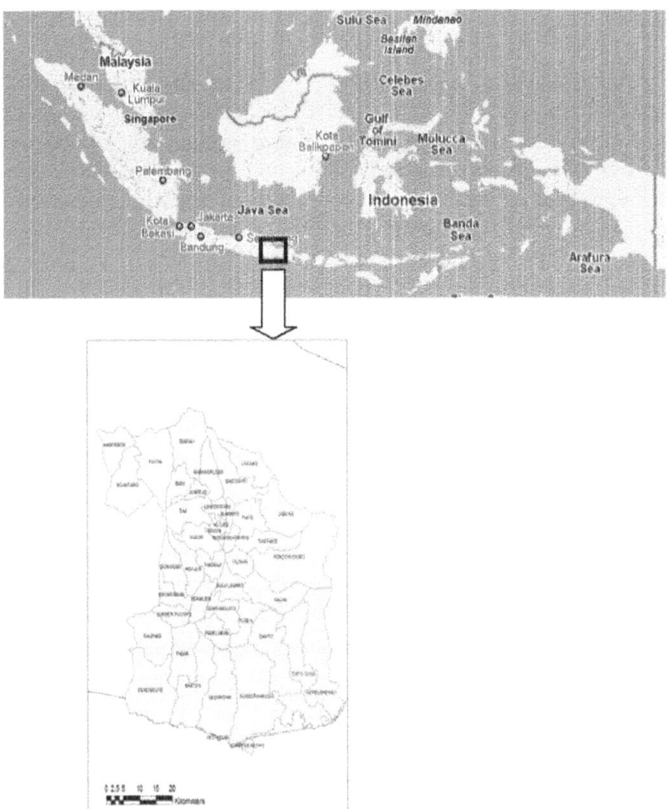

Malang Regency as research areas.

The aims of the research are, first, to identify the characteristics of the land and communities surrounding water springs in the upstream area of the Brantas river basin and, second, to explore public participation in the management of the three water springs with the highest number of users and the highest potential for social conflict.

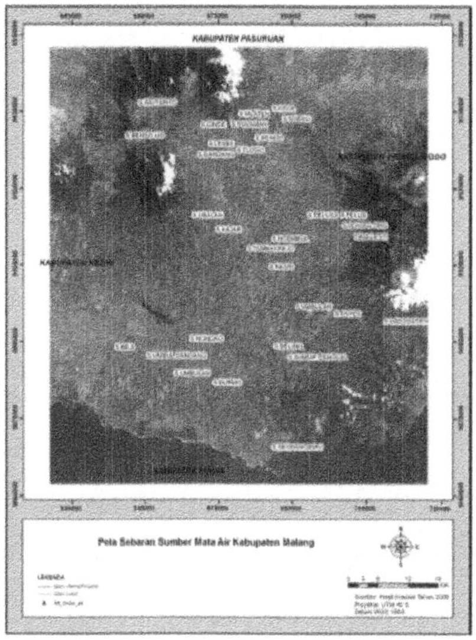

Distribution of 34 water springs in Brantas catchment areas in Malang Regency.

11.2 RESEARCH METHOD

First of all, research activities began with surveying, the purpose of which was to obtain social and biophysical data on 34 water springs, in order to produce a community character profile that would affect the management planning process. Community characteristics were observed (regarding the level of education, livelihood, the level of welfare, etc.). Other social aspects of the 34 water springs were also observed, such as land ownership status, water springs user characteristics, the potential for conflict, the role of the user in conserving water sources, and users' expectations regarding water conservation. The output of the planning model is the availability of the draft water management planning water springs activity and the concrete in the field (three water springs). The expected outcome of the process in Figure 11.3 is the legal establishment of the three water springs management planning treaty and a management forum for water springs recommendations, whose members consist of all the users surrounding the water springs.

11.2.1 Community Profile of 34 Water Springs Areas

Generally, the characteristics of the people surrounding water springs can be described by the level of public education, farmers' livelihoods, and productive age.

Those of productive age in the villages with water springs are expected to contribute, not only by sharing their thoughts, but also by being directly involved in any planning process for water resources, and by participating in the implementation of the pertaining plan toward efforts to conserve the water resource or watershed. Educational level is very influential in

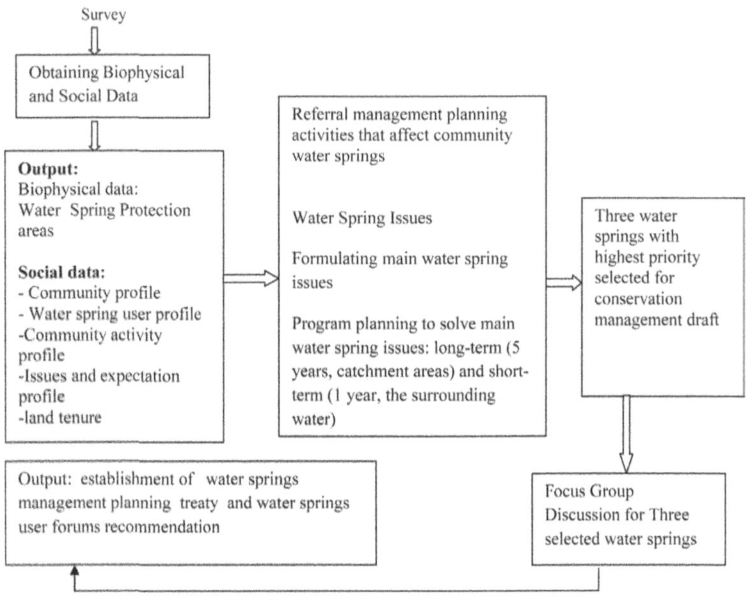

Flowchart of Water Spring Conservation Management research methodology.

improving the quality of human resources. Those with higher education will be able more easily to follow the progress of science and technology. The level of education of most of the surrounding water springs communities is elementary school. Livelihoods analysis is a systematic approach which is still flexible in understanding the situation, access to resources, and disability, options that make the approach useful for the participation of individual and group learning in the community that plays a role that may be new to them. The livelihood of those living in the surrounding water springs is farming, so that dependence on the continuity of natural resources including water is very high (Howes, 2008).

From the level of education, forms of livelihood, age and level of community surrounding the water springs, it appears that efforts have begun to conserve the water springs with a social approach, through the empowerment of communities around the water springs. Involvement of such communities needs to be completed because of their dependence on the availability of water, not only for the basic necessities of life but also for the welfare of the community. Involving communities using participatory techniques helps communities to discover their principal issues and the root causes of emerging problems. Problem solving carried out by the community increases their empathy towards water conservation (Poon *et al.* 2006).

11.2.2 Selection of Water Springs for Priority Management

The opportunities provided by the government district/city governments empower the village to make its life conservation and environmental management including water resources available legally. Lack of village government's role in environmental conservation and management in the region can be caused by a lack of information received, so that the community's potential for knowledge and experience in the field of environmental conservation and management are underdeveloped. The lack of access to such information affects the level of public participation in environmental management and conservation efforts (The USAID, 2009).

The complexity of issues that occur in the field led to the choice of three water springs to be handled with the highest priority:

1. Sumber Awan Water Spring
2. Sumber Umbulan Water Spring
3. Sumber Beling Water Spring.

11.2.3 Focus Group Discussion (FGD) in the Three Chosen Water Springs

Community engagement techniques are used by the FGD. Steps for involvement of water spring users in the preparation of water management planning can be illustrated in Figure 11.4 and it may explain as follows:

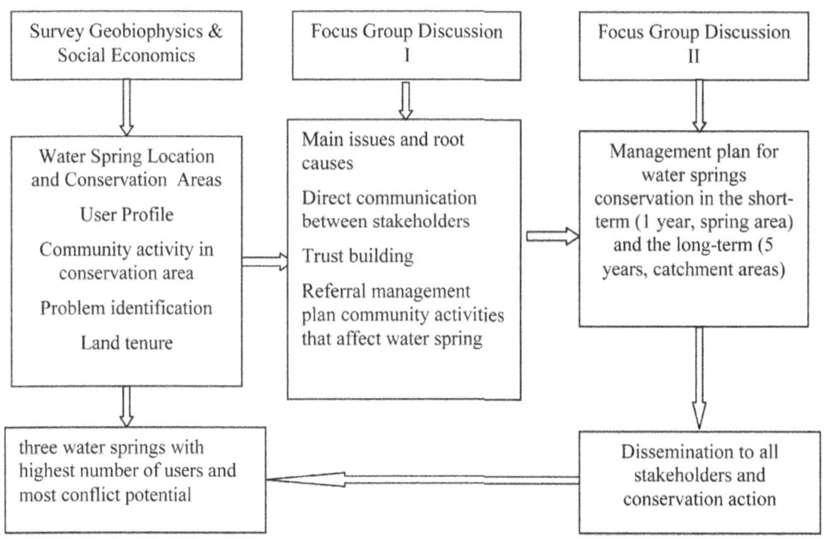

Flow diagram of comprehensive public participatory model.

- Introducing the issues/needs and springs and consciousness; excavating information on the existence of water springs and the surrounding environment, as well as reflection on the reservoirs.
- Identifying the issues faced by each user's water springs, then performing together problem formulation based on priorities related to community needs.
- Identifying problem solving based on the ability of communities and the resources available.
- Implementing problem solving in a concrete action plan for both the short and long term. The planning should include who is doing what and when. Increasingly clear that as the plan is carried out, the huge planning would be done jointly.
- ensuring the action plan will be implemented to organize the user›s water springs by means of forming water springs Forum Communications Manager.

These activities illustraed in Figure 11.5.

Focus group discussion for water spring conservation (Sumber Awan in Singosari district of left photo, Sumber Umbulan in Dampit district of center photo, and Sumber Beling in Sumbermanjing Wetan center of right photo).

11.3 THE OUTCOME OF FGD MODEL

11.3.1 Sumber Awan Water Spring, Toyomerto Village, Singosari District

FGD RESULTS Sumber Awan Water Spring

USER

1. DAM Kab. Malang (Indonesia's Drinking Water Regional corporation)
2. KOSTRAD (Indonesia's Army)
3. BLKI (Indonesia's Job Training Center)
4. YONKAV (Indonesia's Army)
5. AURI (Indonesia's Airforces)
6. Armed (Indonesia's Army)
7. VILLAGE:
   a. Toyomarto Village (5 Drinking Water Users communities Group (HIPPAM) and 3 Water Users for Farming activities group (HIPPA))
   b. Gunungrejo Village (4 HIPPAM and 2 HIPPA)
   c. Village Ardimulyo (HIPPA)
   d. Candirenggo Village (1 HIPPAM and 1 HIPPA)

ISSUES

1. More decreasing in water discharge
2. Declining water quality (bacteriology)
3. Water users' fight between HIPPAM and HIPPA
4. Less maintenance of Regions water spring
5. The lack of public awareness regarding the water spring
6. The lack of a discharge data base
7. Socialization not clear PP (PP No.43 Year 2009 regarding levies and Regulation of forestry minister No. 50 of 2006 on Environmental Services)
8. Irregular operation of Forum Water spring guard
9. Water delivery to the Air Force taken by another institution
10. No water calculation record center for each user

SOLUTIONS

1 Not adding new or additional discharge decision
2 Fixing leaky pipes
3 Doing forestry conservation on catchment areas
4 Changing the cropping system
5 Fixing the leaky pipeline
6 Bringing gutters of household waste water into the fields to increase the irrigation discharge

EXPECTATIONS

1 Maintenance of water springs' high quantity, high quality and continuity
2 User contributions for water used for water conservation
3 Socialization of the regulation of water use, water user associations, and water meter installation.

Circumstances surrounding protected areas near Sumber Awan Water Spring are cause for apprehension, since the number of stands is still lacking, coupled with slopes where one is causing landslides. At a radius of 50 meters, there is some land that is used as a farm (Figure 11.6).

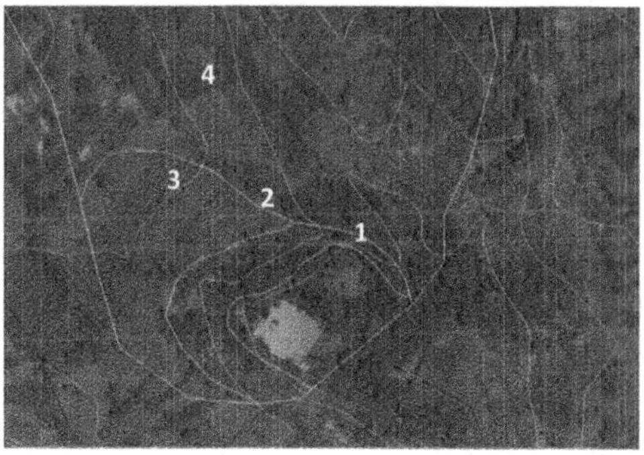

Water spring conservation zone in Sumberawan spring.

Captions.

Zone 1: a complex of water spring discharge. There are at least three large water springs in this zone, other than those which are small and difficult to identify.

Zone 2: a zone of water spring conservation, with steep slopes (25–40%). Visibility on the north side is very minimal and a number of stands in the west are even used for land cultivation.

Zone 3: 50 yards outside the zone but still within a radius of 200 meters, where rainwater will flow into the zone first. Non-irrigated land use in the form of raw water may interfere with the quality of drinking water sources.

Zone 4: Areas' larger catchment is dominated by dry land, the role of the community. Given the interest in this water spring, the community seems enthusiastic enough to conserve water resources with those programs that have been prepared, both by local governments, defense agencies (Air Forces and Army), and the local community (See Figure 11.7).

These recommendation for each functional areas can be explained in Table 11.1.

Conservation action from one Sumber Awan Water Spring User (Indonesian Army).

Recommendations on each functional area.

No	Function	Area (Ha)	Recommendation
1	Protected Area	1.5	Visible in Conservation areas referral mapping
2	Conservation of downstream of water springs	1.5	Tree with anchors and a strong grip
3	Conservation of upstream of water springs	2	terraced springs, enriched trees
4	Catchment area	250	terraced springs, enriched trees

A recommendation that can be given to protected areas is this: Enrich the number stands at zone 2 with plants that have deep roots combined with horizontal roots, such as the plant species shown in Table 11.4

11.3.2 Sumber Umbulan, Umbulrejo Village, District Dampit

FGD RESULTS Sumber Umbulan Water Spring

USER:

1. PDAM Kab. Malang (Indonesia's Drinking Water Regional corporation)
2. Individual water spring users (8 pipes)
3. Agriculture (wetland and dry)

ISSUES

1. A huge volume of the water spring is wasted because there is no good management.
2. There are complaints from the village about compensation.
3. There is still a decision to sell water individually.
4. There is no MOU between the legal owner for land that contains the water spring and the water spring users.

SOLUTION

1. None

EXPECTATIONS

1. Maintaining water quality
2. The legality of water spring users
3. There should be clear water user organizations
4. Water spring management
5. There are rewards for environmental services to improve water resources
6. Sustainable Water Resources
7. For individual water users (organized and constructed)
8. There is a license to use water spring from Indonesia's Department of Energy and Mineral Resources.

Circumstances at the protected area of Sumber Umbulan water spring were not overly concerning, except for the settlements on the ridge flanking the valley where the water spring was located. This is evident in E-Coli bacteria found in considerable amounts in the drinking water reservoir owned by PDAM Kab. Malang. Thus, the main concern is to provide a filter to prevent the bacterium E. coli from contaminating the water spring. At a radius of 50 meters, land use is for production forests. However, some form of mahogany trees have intercropped with crops under the radius, contributing to an increase in substantial erosion (Figure 11.8).

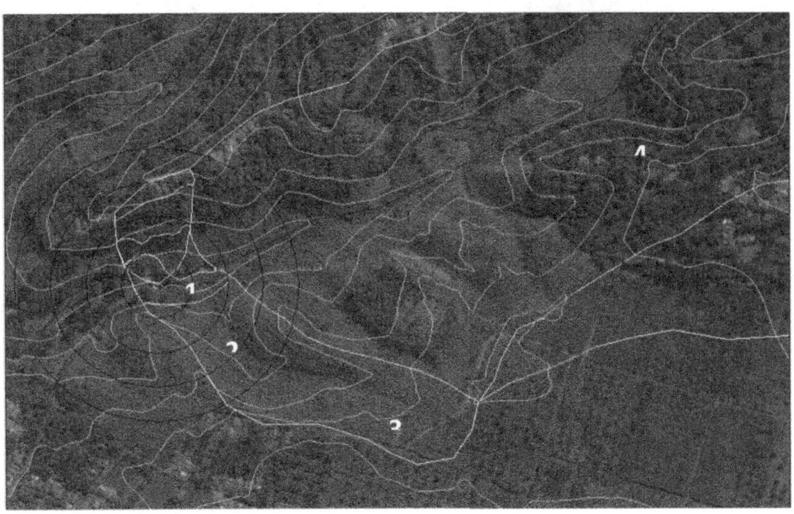

Water spring conservation zone in Sumber Umbulan Water Spring.

Captions.

Zone 1: a complex of water spring discharge. There are dozens of good water spring dots, large or small, that appear on this complex, including in the valley to the upstream and downstream sections (for detail field area may see in Figure 11.9).

Zone 2: a zone of water spring conservation, with steep slopes (25–40%), visible in the north with Mahoni crop intercropping with crops (maize and cassava), while the southern part is Sengon wood. Declining soil fertility appeared under the mahogany tree stands. This decline is felt by local communities.

Zone 3: 50 yards outside the zone but still within a radius of 200 meters, where rainwater will flow into the zone first. Non-irrigated land use occurs in the form of production forests (mahogany and sengon). In the north, on the ridge there are settlements.

Zone 4: catchment area, with land use such as production forests (pine and mahogany), and plant crops: corn, cassava and sugarcane.

These recommendation for each functional areas can be explained in Table 11.2.

Tree planting surrounding Sumber Umbulan Water Spring by local community water spring users.

The community is enthusiastic enough to preserve this water spring, as long as it is authorized to participate in managing it. This is probably because most people already manage their own water needs, drawing from the reservoir and the piping resulting from governmental organizations. A recommendation that can be given to protected areas is this: Enrich the number of stands at zone 2 with plants that have deep roots combined with horizontal roots, such as the plant species shown in Table 11.4.

Recommendations on each functional area.

No	Function	Area (Ha)	Recommendation
1	Protected Area	1.5	Visible in Conservation areas referral mapping.
2	Conservation of downstream of water spring	2	Tree with anchors and a strong grip
3	Conservation of upstream of water spring	5	terraced springs, enriched trees
4	Catchment area	2225	terraced springs, enriched trees

11.3.3 Sumber Beling, Druju Village, District Sumbermanjing Wetan

FGD RESULTS Sumber Beling Water Spring

ISSUES

1. No definite cooperation between PDAM and the villagers
2. PDAM Kab. Malang (Indonesia's Drinking Water Regional corporation) small contribution towards improvements
3. MOU less optimal
4. Untreated water resources
5. Water resource's huge volume of lime content
6. An increase in drinking water customers
7. Environmental management in land ownership rights
8. Pollution from cow waste
9. Meeting the water needs for the citizens about the source of water for citizens around the source, planting greenery revoked.

SOLUTION

1. Finding a legal basis
2. Mutual cooperation between PDAM and the villagers, CSR from PDAM, PDAM carries out greening
3. Carry out monitoring and evaluation activities in the field
4. 6 months of community service (Thanksgiving rituals at water springs every 1 year)
5. Looking for new alternative water resources for adding production capacity
6. There are conservation areas surrounding water springs
7. Making a security source dyke
8. Each resident uses a water pump
9. Replanting efforts surrounding water springs

EXPECTATIONS

1. There are laws or regulations to manage water spring management
2. There are greening activities surrounding water springs
3. a. Publication of clear objectives by one management
   b. Optimalisation in MOU
   c. Water resource conservation
4. Good management
5. Test water quality
6. Inadequate water needs
7. Help seedling planting and plant life assurance
8. Counseling
9. Good coordination for water spring utilities for the public
10. Land acquisition made

The circumstances of the protected areas of Sumber Beling water spring are very unfortunate, because this spring is surrounded by a resident housing complex (Figure 11.10). E Coli bacteria have been found in significant quantities at many locations, such as water reservoir tanks. The main concern is to provide a filter to prevent the bacterium E. coli from entering the water spring.

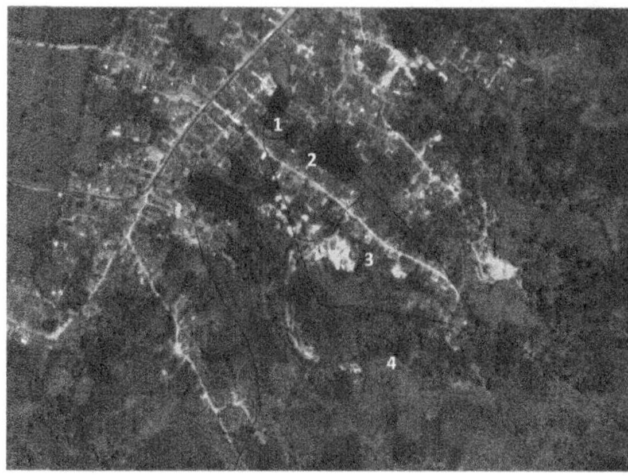

Water spring conservation zone in Sumber Beling Water Spring.

Zone 1: within a radius of 50 meters, on the north and south surrounded by housing residents. The upper water spring has horticulture plants and gardens, while the bottom is a wetland (for detail field area may see in Figure 11.11).

Zone 2: within a radius of 200 meters, is also dominated by a residential population with small gardens.

Zone 3: at the micro-watershed, in which rainwater flows to the location of the water spring and which is still dominated by settlements.

Zone 4: Areas larger catchment in south Malang Escarpment Plato dominated by dry land

These recommendation for each functional areas can be explained in Table 11.3.

Tree planting surrounding Sumber Beling Water Spring by local community water spring users.

Because the water springs are located on individually owned land, so far there have been no conflicts related to water requirements. However, in the future, there may be a prolonged conflict if the water situation is not properly handled. A recommendation that can be given to conserve areas is this: Enrich the number of stands at zone 2 with plants that have deep roots combined with the horizontal roots, such as the plant species shown in Table 11.4.

Recommendations for each functional area.

No	Function	Recommendation
1	Protected area	Planting of vetiver plants to reduce the infiltration of feces, both man and cattle
2	Conservation of downstream of water springs	Enriched trees
3	Conservation of upstream of water springs	Enriched trees
4	Catchment area	Enriched trees

Plantation species that can be planted in Sumber Awan, Sumber Umbulan and Sumber Beling downstream water spring conservation areas.

Index	Weak Anchor, (<0.35)	Medium Anchor, (0.35–0.60)	Strong Anchor, (>0.60)
Weak Grip, (<0.25)	Mango, Sweet Orange	Pecan, Apple	Dadap
Medium Grip, (0.25–0.40)	Muris, calothyrsus, Suren	Acasia, jackfruit, resins, trengguni, tritih dan mindi	Wood Africa, rambutan, guavas, catarina
Strong Grip, (>0.40)	Pinus, Eucaliptus	Durian, keningar, avocado dan Coffee	Randu, Java coffee dan gagar

11.4 WATER SPRINGS MANAGEMENT PLANNING TREATY AND WATER SPRINGS USER FORUMS RECOMMENDATION

In this research, the outcome of FGD from three selected water springs issues were formulated by researchers, who then gave recomendations for The Water Spring Conservation Management Treaty and Water Springs User Forums establishment.

The Water Spring Conservation Management Treaty recommendation contents embody three main items which must agreed by all water spring user stakeholders, namely compensation to landownwers with water springs on their land, contributions by water springs users for the purpose of conserving their water springs areas, and the establishment of a water springs management organization. For detail these recommendation can be seen in Table 11.5.

Details of the Water Springs Management Planning Treaty Recommendation.

LAND OWNER COMPENSATION WHICH HAS WATER SPRING IN THEIR LAND
1. Village Community → Asphalting the road leading to water springs, pipe fitting from water springs into settlement areas, thanksgiving rituals 2. Individual → Building water spring checkpoint post for PDAM Staff
WATER SPRINGS USERS' CONTRIBUTION TO CONSERVE THEIR WATER SPRINGS AREAS
1. Tree Planting 2. All stakeholders work together to clean up water springs periodically 3. Regulatory enforcement of maintaining water springs areas
WATER SPRINGS MANAGEMENT ORGANIZATION
PDAM Owner → PDAM together with all stakeholders in water spring user forums Non-PDAM Owner → Water spring user forums

Government affairs responsibilities that can be handed over to villages.

Official	Activities
Agriculture	Regulation of water spring users for farming activities Counseling of water spring users in farmer groups
Mines and Energy and Mineral Resources	Recommendations for permits for the utilization of ground water Recommendations for licensing and the utilization of ground water or springs in the village
Plantation and Forestry	Conservation
Spatial Planning	Environmental and rural settlements arrangement
Public Work	Management and utilization of existing water projects in villages Management of water resources in the village
Financial	Rural financial balance management from certain levies in districts
National Land Agencies	Giving certificates of land rights Structuring and mapping of land use
Planning and regional development agencies	Preparation of participatory village development planning
Social	Community participation improvement in rural development

The recommendation of this research is for the government to give Water Spring User Forums established by legal law, depending on the level of needs and development of local communities. The main role of these forums is to manage communities' water springs areas. Furthermore, for the sake of the preservation of water springs, in preparing a program based on the direct involvement of water spring users, it is important to ensure the sustainability of the program. For detail these recommendation can be seen in Table 11.6.

11.5 CONCLUSION

The preparation of a management plan directly involving water springs users seems to have a positive impact on water springs conservation programs. It can be concluded that water springs users and the community feel that such direct planning is in accordance with their needs, creates a high sense of ownership of the program, and reduces dependence on outside parties to take the initiative in formulating programs. In the future, hopefully this prototype model can be implemented globally, especially in all areas of Indonesia.

11.6 ACKNOWLEDGEMENT

This research regarded Refferal Water Springs Protection Zone program, which was supported by Environmental Service Programmes (ESP) East Java Regional–USAID and Indonesia's Drinking Water Regional Bureau of Malang Regency (PDAM Kab. Malang) in 2009.

11.7 REFERENCES

Hakli J. and Minca C. (2009). *Social Capital and Urban Networks of Trust*. Ashgate, England.

Hammer M. J. and Hammer Jr. M. J. (2004). *Water & Wastewater Technology, 5th Ed.*, Pearson, New Jersey.

Howes H. (2008). *Strategic Planning for Water*. Taylor & Francis, London.

Poon J. P. H., Button K., Nijkamp P., (2006). *Social Planning*, Elgar, The UK.

The Indonesia's Environmental Ministry Decree No. 32 Year 2009, The Conservation and Management of the Environment, The Government of the Republic of Indonesia.

The Law No. layout. 26, 2007, Indonesia's Spatial Planning, The Government of the Republic of Indonesia.

The USAID (2009). *Refferal Water Springs Protection Zone in East Java report*, Indonesia.

The Water Resources Act No.7, 2004, Indonesia's Water Resources, The Government of the Republic of Indonesia.

12 Network analysis of community based water management

Ismu Rini Dwi Ari[1], Hayeong Jeong[2], Kakuya Matsushima[3], Kiyoshi Kobayashi[4]

[1] Engineering Faculty, University of Brawijaya, MT. Haryono 167 Malang, 65145, Indonesia. E-mail: dwiari@ub.ac.id

[2] Graduate School of Engineering, Kyoto University, Katsura, Nishikyo-ku, Kyoto, 615-8540, Japan. E-mail: jeong.hayeong.2z@kyoto-u.ac.jp

[3] Graduate School of Engineering, Kyoto University, Katsura, Nishikyo-ku, Kyoto, 615-8540, Japan. E-mail: matsushima.kakuya.7u@kyoto-u.ac.jp

[4] Graduate School of Engineering, Kyoto University, Katsura, Nishikyo-ku, Kyoto, 615-8540, Japan. E-mail: kobayashi.kiyoshi.6n@kyoto-u.ac.jp

Abstract

Lack of access to safe drinking water is a world wide problems, whereby one of the cause is inappropriate water governance. Nevertheless, there is an existency of community based water management that able to mobilize local resources on connecting piped water supply system for their members. This voluntary association plays a role as one of strong alternatives to existing water supply system by the local government. This research proposes Social Network Analysis approach to investigate social structure of the community through three indices of centrality with respect to empirical case of two villages in Singosari district Malang regency Indonesia. The result of research demonstrates three types of social structure of the community which is their social tie might bring them to have ability to maintain their community based water supply system in a specific way. In addition, this research may provide evidence the linkage of collective action theories and the social capital approach that choice to join community based water supply system may reflect their individual choice as well as collective action.

Keywords: centrality, community based water management, multiple memberships, network analysis, social structure

12.1 INTRODUCTION

One of the most indispensable human necessities is the availability of access to safe drinking water. WHO (2010) stated that more than eight out of ten rural inhabitants live without access to improved drinking-water sources; moreover, according to WHO and UNICEF (2006), the number of urban residents without adequate water services is increasing rapidly. According to the report of the Asian Water Development Outlook (AWDO), future water crises in Asian countries will result not from the physical scarcity of water, but from inadequate or inappropriate water governance, including management practices, institutional arrangements, and socio-political conditions, which leave much to be desired (Bird *et al.* 2008). Estimates of access to safe drinking water are a cornerstone of most international assessments of progress toward solving global and regional water problems, particularly to reach target 10 of the Millennium Development Goals (MDGs) aiming to reduce by half the proportion of people without sustainable access to safe drinking water by 2015.

In the case of Indonesia, among a total of 231.6 million inhabitants, only around 49.7% of the citizens have access to water (Statistic Centre Board Indonesia, 2005). Almost 40% of citizens have access to a piped water supply system from a local drinking water company called PDAM, which belongs to the public sector and whose main task is to distribute drinking water to all citizens in the city or regency area. Meanwhile, around 10% of citizens have access to a piped water supply system managed by the local community, generally in rural areas. Utilizing a nearby water source, a group of community members collaborates to deliver piped water to their members. A collective action that may be defined as action on the part of one or more people striving to achieve the objectives or satisfy the common interests of the group, implies devising frameworks that limit the pursuit of individual self-interest and sustain the benefit shared by the group. In this sense, it may be assumed that the community-managed water supply system is a strong alternative to the public sector's existing water supply system.

Focusing on respondents who have the option of accessing a piped clean water supply system from HIPPAM and PDAM, this research aims to scrutinize the social structure of the residents through memberships in community groups from which affiliation data are collected as they refer to respondent-by-community group participation. The collection was conducted in order to understand social relations among the respondents, specifically whether respondents tended to have social relations primarily within their own social classes. Furthermore, the purpose of the data collection is to investigate how the structure of respondents' connectivity affects their choice of access to a piped clean water supply system. This research proposes a Social Network Analysis approach to investigate the social structure of the community through three indices of centrality with respect to the empirical cases of two villages in the Singosari district of Malang Regency, Indonesia. Moreover, this research focuses upon affiliation networks representing the affiliation of a set of respondents with respect to a set of events, on the basis of shared members of community groups in the local area. Here it associates social capital with local level voluntary associations in which individuals jointly participate in collectivities of community groups that exist in their neighborhood. This research illustrates evidence that strong social ties might give communities the ability to mobilize their local resources to achieve common goals, as mentioned by Putnam (2000): "a well-connected community – the one with 'community social capital' will have better ability to mobilize local and extra-local resources to effectively act."

This entire chapter consists of four sections, beginning with section one, which contains a description of some of the basic ideas covered here. Section two discusses the research methods covering social network analysis, focusing upon centrality, determination of sample size, and multiple memberships in community groups. Then, section three presents results and discusses the research, with attention to two main points. Firstly, it applies the methods of Social Network Analysis with three indices of centrality measure to scrutinize social structure in the community that possesses the ability to develop a community-managed water supply system. Secondly, in order to give broader perpectives about social networks among villagers, it continues with a portrayal of trustworthiness among neighborhoods and their perspective toward the village living environment. Last, this paper will be finalized with a conclusion in section four.

12.2 RESEARCH METHODS

12.2.1 Social Network Analysis: Centrality

According to Wasserman and Faust (1994), the notion of centrality is to identify the most important actors in a social network, located in strategic locations within the network. This means that the actor with centrality is the most active and has the most ties to other actors in the network. In other words, an actor is considered prominent if the ties of the actor make the actor particularly visible to other actors in the network. Centrality measures are used to understand group structure, that is, which actors have most access and are most central in the network. Three indices to quantify the prominence of an individual respondent embedded in a network are (i) degree, (ii) closeness, and (iii) betweenness. In addition, the index of centralization is between 0 and 1, whereby the very active actor has a maximal central index.

Referring to Wasserman and Faust (1994), this research scrutinizes affiliation network as a nondirectional relation in which the *i*th row of the matrix X, $(x_{i1}, x_{i2}, \ldots, x_{ig})$, is identical to the *i*th column $(x_{1i}, x_{2i}, \ldots, x_{gi})$. The prominence of actor *i* within a network is based on the pattern of g – 1 possible ties in the affiliation network matrix. Thus, for nondirectional relation, a central actor is defined as one involved in many ties, regardless of the reception (being the recipient) as well as the transmission (being the source) of many ties.

12.2.1.1 *Bipartite degree centrality*

According to Everett and Borgatti (2005), in the two-mode network, the raw degree centrality for an actor n_i is defined as the number of events participated in by the actor, $C_D(n_i)$.

Modified from Everett and Borgatti (2005), the two-mode normalization of degree centrality for an actor is represented by the following formula. The *h* is defined as the number or the size of events, *M*. In addition, this is a nonlinear whereby the two-mode normalization towards actors could be scaled in a different way depending on the size of actors.

$$\text{Actor } n_i \text{ normalized centrality} = \frac{C_D(n_i)}{h} \tag{12.1}$$

Referring to Wasserman and Faust (1994), isolated actors in particular, $d(n_i) = 0$, may be eliminated from the network with no effect on the current ties of actors. As a consequence, higher actor normalized degree centrality will be formed, since the number of the denominator in standardized indices is reduced, but the order level of centrality will be unchanged.

12.2.1.2 *One-mode betweenness centrality*

The notion of betweenness centrality is that an actor is central if it lies between other actors on their geodesics, implying that to have a large "betweenness" centrality, the actor must be between many of the actors via their geodesic. In this sense, the actors who lie on the many shortest paths between the pairs of other actors – these "other actors" potentially might have some control over the interactions between the two non-adjacent actors (Wasserman and Faust, 1994). Furthermore, Wasserman and Faust (1994) argue that actors in the middle, the ones between the others, play important roles in the network; their betweenness may cause them to have more interpersonal power on the others.

In order to quantify the notion of betweenness, firstly Wasserman and Faust (1994) consider the probability of a communication flow, by proposing two assumptions: (i) lines of a path whereby communications will travel along the shortest route to reach actor *k* from actor *j* have equal weight, so that (ii) such that a communication follows one of the geodesics. Then, as Freeman (1977) estimated that the probability of a communication from actor *j* to actor *k* takes a particular route is denoted as g_{jk} be the number of geodesics linking the two actors. So that, the probability of the communication using any one of the chosen paths is inverse g_{jk}. Secondly, they consider the probability that a different actor, *i*, is "involved" in the communication between the two actors in which $g_{jk}(n_i)$ is the number of geodesics linking the two actors that contain actor *i*. Thus, in line with the estimation of Freeman (1977), the actor betweenness index for n_i is the sum of these estimated probabilities over all pairs of actors not including the *i*th actor for *i* distinct from *j* and *k*.

$$C_B(n_i) = \sum_{j<k} g_{jk}(n_i)/g_{jk} \tag{12.2}$$

Betweenness of actor centrality is a sum of probabilities that a minimum value is zero when n_i falls on no geodesics. And then, the number of pairs of actors not including n_i is the maximum value that is $(g-1)(g-2)/2$. Thus, the value of actor betweenness is between 0 and 1, as formulated in the following model.

$$C_B'(n_i) = C_B(n_i)/\left[(g-1)(g-2)/2\right] \tag{12.3}$$

Just as with degree centrality, the betweenness indices can be computed even if the graph is not connected or is different from closeness indices that it has to be reachable to actors in the network.

12.2.1.3 One-mode closeness centrality

According to Wasserman and Faust (1994), actor closeness centrality measures how close "the distance" an actor is to all other actors in the network. An actor has the position of closeness centrality if it can quickly interact with all others, so that centrality is inversely related to distance. This means that increasing geodesics in length between actors decreases the centrality of the actors.

Following Freeman (1979), Wasserman and Faust (1994) defined the simple measure for closeness centrality as a function of geodesic distance as proposed by Sabidussi (1966). The distance between actors *i* and *j*, denoted as $d(n_i, n_j)$ is the number of lines in the geodesic linking actors *i* and *j*, as a distance function, and it is the length of any shortest path between the actors. Hence, the total distance that *i* is from all other actors is $\sum_{j=1}^{g} d(n_i, n_j)$, in which the sum is taken over all $j \neq i$. Thus, the index of actor closeness as defined by Sabidussi (1966) is

$$C_C(n_i) = \left[\sum_{j=1}^{g} d(n_i, n_j)\right]^{-1} \tag{12.4}$$

In order to make comparisons of values across networks of different sizes, standardized closeness centrality is expressed as follows (Wasserman and Faust, 1994).

$$C_C'(n_i) = \frac{g-1}{\left[\sum_{j=1}^{g} d(n_i, n_j)\right]}(g-1)C_C(n_i) \tag{12.5}$$

The normalized index ranges between 0 and 1, and can be viewed as the inverse average distance between actor i and and all the other actors (Wasserman and Faust, 1994).

In line with previous researchers, Faust (1997) also defined closeness centrality as based on geodesic distances - shortest paths of the central actors from other actors in the network, applicable only to dichotomous relations. Moreover, she mentioned that in cases where some pairs of actors have no link between them, then the distance between them will be infinite and the closeness centrality will be undefined. In other words, as defined by Wasserman and Faust (1994), the geodesic distances of the isolate actor, n_k – with degree 0 because the actor is not reachable from any other actors, are infinitely long, $d(n_i, n_k) = \infty$ for all $i \neq k$.

Referring to Brandes and Pich (2006), closeness centrality and betweenness centrality are based on shortest-path distances (geodesic distances). It is the number of links in the shortest path between actors. Hence, any kind of influence or communication by one actor on another declines with the distance between them. As a consequence, an actor has high closeness centrality when the total (and therefore also average) distance of the actor to all other actors is small; meanwhile, an actor has high betweenness centrality when the actor is located in relatively many shortest paths connecting pairs of others.

12.2.2 Determination of Sample Size

The impetus for this chapter was the belief that a piped clean water supply connection to houses is the most appropriate water supply system by which people may live with safe water. In this area of study, there are two types of piped clean water supply connections to houses provided by PDAM and HIPPAM. PDAM is a formal local drinking water company at the level of city or regency. In the research area, water supply services are maintained by PDAM Unit Branch, Singosari district, Malang Regency. And HIPPAM is an abbreviation for a residents' association of drinking water users, a voluntary association of villagers which creates and manages a piped water supply system within the scope of the local community. In the research area, there are six HIPPAM groups in Toyomarto village and one HIPPAM group in Candi Renggo village (Figure 12.1).

A sample of which are defined as follows:

1. The respondent is currently a HIPPAM member or a PDAM customer getting access to a piped water supply system;
2. The respondent lives in a hamlet where currently water services from HIPPAM and PDAM are available as represented by the geographic distance of each respondent's house from the main pipeline of HIPPAM and PDAM.

Among 7 hamlets in Toyomarto village, the water pipeline network of PDAM and HIPPAM is available to residents of Sumberawan and Ngujung hamlet, meanwhile among 15 hamlets in Candi Renggo village, the water pipeline network of PDAM and HIPPAM is available to residents of RW 6. Among 436 households in Sumberawan hamlet, except for majority households with joint participation in HIPPAM groups, there are some residents who still acquire water from the river, though the exact number is hard to determine and none have become PDAM customers. Residents of Ngujung hamlet (424 households) get access to water from PDAM, HIPPAM, individual wells and public hydrants. And residents of RW 6 have options to acquire water services from both PDAM and HIPPAM.

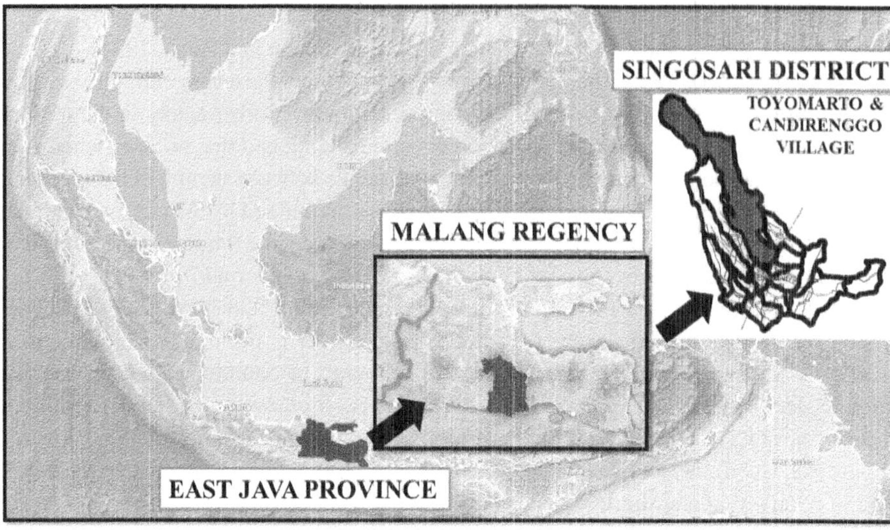

Figure 12.1 Research area.

This research attempts to measure the social network of a set of respondents and a set of voluntary organizations, and to examine the relations that form a collection of ties when respondents join voluntary organizations. So that, primary data on this paper, as mentioned by Narayan and Pricehett (1999), employed an index of social capital at the household and community level, which included characteristics of informal and formal groups and networks to which people belonged. In this respect, the respondents were asked to state in which community groups they participated. Since this research will measure the affiliation relations of a set of respondents in joint participation in a set of community groups which are kinds of voluntary associations existing at the community level, the research may include in the analysis some particular community groups which have characteristics as determined by the following scholars. First, there is the classification of voluntary organizations by Tao (1999) into three categories – (i) self-help group, (ii) service delivery providers, and (iii) campaign group. Second, the three characteristics of the community as a group, defined by Aoki (2009), are that (i) membership is relatively stable in that members share (limited) information or the use of certain local commons, (ii) members create internal rules as substantial regulations or are able to form their community norms as endogenous rules, and (iii) they utilize these community norms as self-enforceable without third-party involvement. Third, following Wasserman and Faust (1994), certain community groups, such as those which no longer exist due to a lack of participants or those to which only one respondent belongs, cannot serve to illustrate social interaction among members and therefore may be excluded from consideration. Thus, this research scrutinizes the social structure of the 234 households in the research area in affiliation networks of collectivities linked through the 4 community groups, as seen in the data presented in Table 12.1.

The type of social activity that a majority of respondents engage in is the religious group, which encompasses a collection of residents who voluntarily establish a community group whose main purpose is to deepen understanding of religion. Since the religion of most residents is Islam, then in general, they have weekly meetings for recitation of the Qur'an, usually conducted on Thursday night after Isha prayer, held alternately in the homes of group members. Periodically, residents also organize special meetings where they learn together how to care for the Moslem corpse. This is an informal community association whereby structurally the organization consists of the head, secretary and treasurer of the group, voluntarily chosen from among members to serve for a certain period. Then, together they develop community norms, determining, for example, the committee members for the next year, the order of house rotation for their weekly activities, ways to encourage passive members, and the like. Referring to Tao's classification (1999), this religious group may be similar in character to a self-help volunteer group.

Table 12.1 Sample respondent based on memberships in community groups.

		Religius	**Cultural/ Social**	**Community Org.**	**Finance**	**Total Respondent**
Toyomarto	Sumberawan Hamlet					100
Village	HIPPAM	86	12	11	11	
	Ngujung Hamlet					64
	HIPPAM	42	13	4	3	
	PDAM	16	2	1	1	
Candi	RW 6 Hamlet					70
Renggo	HIPPAM	45	7	7	5	
Village	PDAM	1	0	0	0	
Total sample						234

There are 34 respondents who participate in cultural/social group. The cultural/social group that they join is a kind of regular social gathering for the purpose of conducting a lottery called *arisan* in the Indonesian language. As with a religious group, the structure of this organization also consists of the head, secretary and treasurer of the group, voluntarily chosen from among members for a certain period. This voluntary association not only has the social domain of the religious group, but it has also has the economic domain, which is mutually beneficial to its members. Moreover, compared to the previous group, this voluntary association has more stringent rules in order to run a steady and sustained organization for a certain period. For instance, a social gathering group consists of 35 members with a regular meeting once a month to conduct a money lottery. Then, if the members agree to having one winner at each meeting, then around 3 years will be needed to finish one round of lotteries, which will allow each member a turn to win the lottery. The money used in the lottery is comprised of cash contributions from each member, which is paid at the beginning of the meeting. Therefore, to finish completely one round, they need a strong commitment from each member to participate for the whole period in a fair manner. This means that the members need to compose their endogenous rules to secure their membership in order to govern their voluntary association properly regarding such matters as how to determine the capital to be used in the lottery, how to determine the winner for each round, how to handle members who have difficulty paying dues, and the like. In addition, the capital used for the lottery is not limited only to money, but it is also possible to use certain daily goods, for example a number of eggs, an amount of sugar, a liter of cooking oil, or the like. And as with the earlier group, referring to Tao's classification (1999), this social gathering group may be close in character to a self-help voluntary group.

There are 23 respondents who participate in a community organization group, the most common type of which is a kind of cooperative organization, referred to as *koperasi* in the Indonesian language. In accordance with UUD (Indonesian Constitution) 1945 Section 33, a cooperative is considered a business organization if the cooperative has principals and basic values relevant to the characteristics of Indonesia. Referring to Tao's classification (1999) this community group could be categorized as a service delivery provider, whereby the purpose of the group is to provide service to people who need it. For instance, a farming cooperative will provide services to support agricultural activity such as providing qualified paddy seeds, fertilizer and the like. In another case, the cooperative will arrange to market products of the members, such as sandals as handicrafts of the villagers. Additionally, the structural organization of this community group is more established than for the other three groups, mostly in the form of incorporated organizations.

There are 20 respondents who participate in a kind of community 'finance' group. This is a type of community microcredit based on traditional informal groups, in which the groups provide microloans for their members or nonmembers who need financial support from their savings and loan voluntary associations. This voluntary association usually consists of the head of the organization, a secretary and a treasurer elected by and from among the members to serve for a particular period. All the rules and norms are set up entirely by the group members with no intervention from a third party. In this sense, group members determine their own rules, such as the amount of money each member must deposit to the association to form joint capital, the maximum amount of money that can be borrowed by a member, the interest rate for members, the rules for benefit sharing among members, the rules for members late in paying their debts, and the like. Moreover, in case the group's capital is more than enough to provide services for members, group members determine the rules regarding nonmembers who might want to borrow money from the group, and the like. Referring to Tao's classification (1999) this community group could be categorized as a service delivery provider.

Referring to the Instruments of the Social Capital Assessment Tools developed by the World Bank (Christian *et al.* 2004), this research continues to scrutinize two more aspects of social networks in order to capture more comprehensive understanding about the social structure of the community in the research area. The first concerns trustworthiness among residents with regard to lending and borrowing, and the level of trust within three years. The second, which covers thirteen items, concerns the perspective of the community towards their living environment.

12.3 RESULTS AND DISCUSSINS

12.3.1 Chronological Development of Water Supply System

There are three natural springs as the main water resources in the research area. First is Candi Sumberawan Spring, which lies at an altitude of 671 meters above sea level in Sumberawan hamlet, Toyomarto village, Singosari district. This spring plays a very important role in supplying clean water to the residents of Singosari district, covering several hamlets in Toyomarto village, such as Sumberawan and Ngujung. In addition, its water has been utilized by residents for domestic purposes (drinking and cooking, and also bathing and cleaning), as well as for irrigation for their paddy land.

The second natural spring employed by the residents is Sumber Banyon, which lies at a higher altitude than Candi Sumberawan Spring but has a smaller capacity. Particularly, this spring is utilized by HIPPAM Sumber Banyon for domestic purposes, such as for drinking water for some households in Sumberawan hamlet. Thirdly is Ken Dedes Pond, in RW 6 Candi Renggo village, a natural pond that serves as a local recreational swimming pool managed by the Indonesian Air Force in Singosari district. The pond also supplies piped drinking water for the Indonesian Air Force at their base camp in Pakis district – the neighboring district of Singosari district. Next, water usage is shared with local residents in RW 6 through a cooperative association called Primer Koperasi Tirta Perwita.

12.3.1.1 Water supply system development in Toyomarto village

The inception of a piped clean water system was introduced by two types of agents, namely the Indonesian Military and BLKI (Industrial Education Training Centre) Singosari. Divided into three periods, they constructed an iron water pipeline to utilize Candi Sumberawan Spring to supply their own territory exclusively, without any water connection for the local residents.

First, a piped water system in Toyomarto village was developed in 1961 by KOSTRAD (Army Strategic Reserve Command), which constructed a water pipeline to utilize Candi Sumberawan Spring to supply their military base in Ardimulyo village. Second, in the period of 1962 – 1963, BLKI (Industrial Education Training Centre) Singosari developed another water pipeline also utilizing Candi Sumberawan Spring to supply their education training centre in Ardimulyo village. Third, gradually from 1962 to the 1980s other water pipelines were constructed by two divisions of the Indonesian Army and Indonesian Air Force in Singosari district. They also utilize Candi Sumberawan Spring and transmit water to the military base of YON ARMED (Army Field Artillery Battalion) in Candi Renggo village and YONKAV (Army Cavalry Battalion) in Ardimulyo village. For the Indonesian Air Force, water was transmitted to their military base in Pakis district, which neighbors Singosari district. So far, there are no available recorded data for water capacity exploited in these iron pipes. Foremost, the entire water pipeline has been developed exclusively to supply water for the five exclusive territories outside Toyomarto village.

For more than two decades, from the 1960s to the 1980s, though villagers could see all these piped water supply systems, they still had to fetch water for their daily needs directly from Candi Sumberawan Spring using jerricans or plastic buckets. Villagers use Sumberawan River and Glatik River as their public bath wherein they bathe and clean their clothes, their cattle, and their motor bikes/trucks. Moreover, spillover water from the spring is also utilized by farmers for irrigation purposes. The spring and the river function as common pool resources (CPRs) for the villagers. CPRs are characterized by the rivalry for their consumption and the difficulty or even impossibility of excluding anyone from utilizing them. Tradionally, water resources have been maintained for the people's needs – among villagers and for different needs – based on local wisdom, where residents live harmoniously and respect their natural resources as part of their entire life. But, it seems extremely hard for villagers to compete with the powers who have brought the spring water outside the village.

The second step was in 1984, when PDAM Branch Unit Singosari built an open water storage unit around 20 × 20 square meters in size, connected with an iron transmission pipeline at a diameter of 200 – 250 millimeters. This pipeline network transmits water from Candi Sumberawan Spring with a capacity of 10 liters per second to a main water reservoir in Candi Renggo village. At the inception of the construction work, PDAM offered to residents in Toyomarto village, particularly the Sumberawan hamlet community, to install piped drinking water connections to their houses, since the spring lay within their hamlets. In fact, none of the residents in Sumberawan hamlet became PDAM customers. The cause of this rejection was unclear. Either it was related to residents' traditional way of utilizing water from springs or rivers, or it was related to the cost that the residents would have had to bear for the physical installation of distribution pipelines to their houses, as well as the unfamiliar monthly charge, since traditionally they had always acquired water from springs or rivers at no cost. Nevertheless, this episode was the residents' first introduction to a piped water supply system (Figure 12.2).

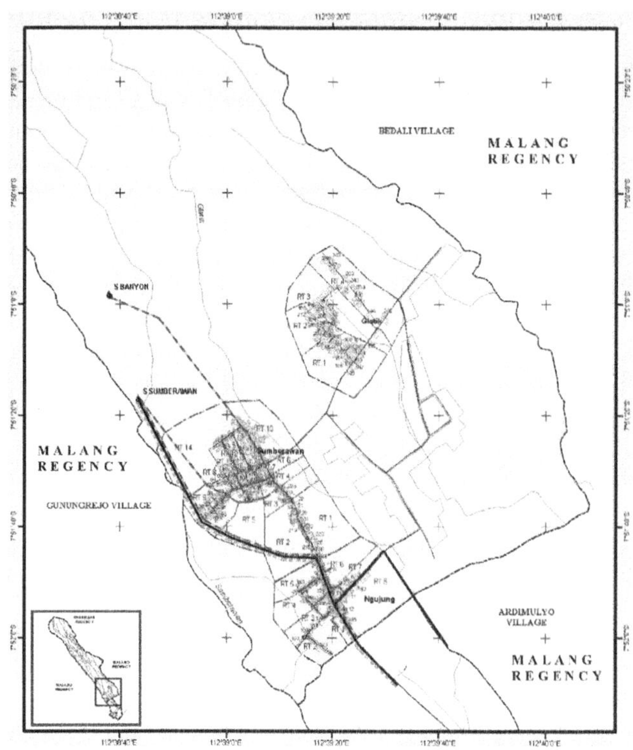

Figure 12.2 Piped water supply system in Toyomarto village. Note: The blue solid line is the main pipeline of PDAM, the magenta dash line is the main pipeline of HIPPAM, and the green dash line is the main pipeline of the Indonesian Military and Education Institution

University students in Indonesia must participate in a community service activity as a requirement for graduation with a bachelor's degree. In 1996, a group of students from Brawijaya University initiated construction of the first public hydrant in Toyomarto village called HIDRAM in the vicinity area of Candi Sumberawan spring. This communal facility had been utilized by residents in the upper area of Sumberawan hamlet until 1996. In this sense, for the first time, villagers were introduced to another form of CPRs – a public hydrant as a communal public infrastructure that brought water closer to their house than what they had had to carry in jerricans or plastic buckets directly from the spring.

In the tranquility period, the Military of Indonesia (ABRI) conducted a social responsibility program whereby together with residents they would develop some basic public infrastructures in order to improve the quality of life in particular marginal areas thoughout Indonesian territory. Between 1991 and 1992, a group of ABRI conducted their social responsibility program in Toyomarto and Candi Renggo village. One of their activities was constructing a new water transmission pipeline network from Candi Sumberawan Spring to Ngujung hamlet, whereby the pipeline network followed the earlier pipeline of PDAM. Pipeline from the spring was developed along the river in the southwest of Sumberawan hamlet and transmitted water to a main water reservoir in Ngujung hamlet. This facility became the first public hydrant for residents in the vicinity area who fetched it using plastic buckets or jerricans for domestic purposes. Then, this public hydrant motivated residents in Ngujung hamlet to develop piped clean water connections to houses and to establish their first and only HIPPAM group in the hamlet, called HIPPAM Air Minum Bersaudara.

Learning from their neighbor in Ngujung hamlet, and driven by the head of hamlet (Mr. Suwandi) and one informal leader (the deceased Mr. H. Ashari, a religious figure) around 1994, residents in Sumberawan hamlet started to construct their own piped clean water supply system with connections to houses, as well as to develop some public hydrants. Then, over time, as the number of users increased, in 1996 the community leaders initiated a grouping of the users into 4 HIPPAM groups, namely Gotong Royong I – IV. Around 1996, residents in the upper part of Sumberawan hamlet, now used to fetching water from HIDRAM, also initiated the establishment of another HIPPAM group called HIPPAM Sumber Banyon, named for the spring they utilized. Then, in 1991, residents in Sumberawan hamlet received financial support from the government of Malang regency to improve the pipeline network. The growth of the community-managed water supply system in Sumberawan hamlet seems to be an effect of the flow of information among residents, driven by the leader of the hamlet, enabling members to achieve their common goal of having piped access to water.

In fact, however, the pipeline system is connected to only a very few residents – around 15% of all villagers, as reported in the Data Base of Clean Water in Malang Regency (2008). In other words, a majority of residents still have to fetch water from springs a certain distance from their houses.

12.3.1.2 Water supply system development in Candi Renggo village

Around the 1950s, the Indonesian Air Force in Singosari district had water rights to manage Ken Dedes Pond. There had been three main purposes of water usage. First, the pond served as a recreation area with a public swimming pool. Second, the institution utilized the pond to supply water to their military bases in Pakis district exclusively, constructing a piped water network with a capacity of 20 liters per second. But, at present, due to water leakage resulting from very old pipelines and free riding by some residents living along the pipeline, this institution pipes water at only around 5 liters per second. This situation has triggered a conflict between the military and the residents, particularly as related to newly constructed buildings whose location above the pipeline hinders repair activity. Thirdly, the institution has given a water share to the community nearby for drinking as well as for irrigation, under the management of cooperative Primer Koperasi Tirta Perwita.

In 1984, PDAM Branch Unit Singosari constructed main piped drinking water utilizing water from Sumberawan Spring at a capacity of 10 liters per second. The water has been transmitted to a main water reservoir in Candi Renggo village using iron pipe at a diameter of around 200 – 250 millimeters (Figure 12.3). The main pipeline was constructed along the village road and served as a piped water connection to houses of residents along the road, including the RW 6 hamlet.

Around 1989, a new residential area, currently RW 12, was developed. This was a specific residential area provided for members of the Indonesian Military, particularly the Army. There was an agreement that in first 3 years the residents would have to acquire water from their personal wells, then after completion of the pipeline construction, they would get access to the PDAM piped water supply system. This is the main reason for the current situation whereby a majority of residents have acquired connection to clean water services from PDAM, while the rest continue to acquire clean water from their individual wells. Moreover, some residents utilize both water sources to get a more sufficient water supply due to certain limitations of PDAM water services.

There is a lot of fertile land in RW 6 that is utilized by villagers for rice paddies. The inception of community-managed water systems is based on the need for villagers to have equal shares of irrigation water for their rice paddy fields. This is an initiation of the establishment of a kind of agricultural cooperative for farmers. Over time, since in the village there is no formal agricultural cooperative capable of adequately supporting agriculture

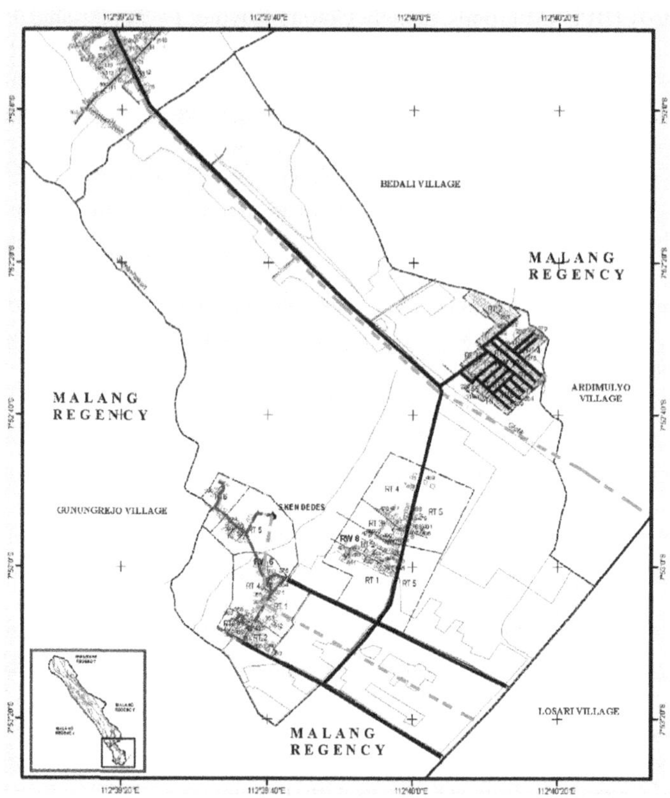

Figure 12.3 Piped water supply system in Candi Renggo village. Note: The blue solid line is the main pipeline of PDAM, the magenta dash line is the main pipeline of HIPPAM, and the green dash line is the main pipeline of the Indonesian Military and Education Institution

through such activities as providing good paddy seeds, fertilizer, and the like, the residents took the initiative to establish a voluntary association for the purpose. Then in 1998, this association became an incorporated cooperative institution called the Primer Koperasi Tirta Perwita, which now provides members with products supporting agricultural activities as well as with agricultural microcredit. At the same time, a group of people also formed a voluntary community water association to provide piped water supply system for domestic purposes. Then, in 2006, when the organizational structure of the cooperative became stronger, it was decided to merge the two main activities into one organization in order to be able to give more sufficient and effective services to the members. All these things were smoothly accomplished because of the good relationship between residents and the Indonesian Air Force. On one side, there is a cohesive relationship among residents in RW 6 that matched with the idea of the Indonesian Air Force to share their water usage with the residents in the vicinity of the pond. This mutual collaboration may have a good impact on both sides. However, needless to say, with regard to the piped water supply network, there remains the problem of free riding by a few residents along the pipeline.

Though development of the water supply system in Candi Renggo village has been underway for around 5 decades, in fact the pipeline water supply system has been connected to only a very few residents – at around 35% of the total villagers as reported in the Data Base of Clean Water in Malang Regency (2008). This is slightly better than conditions in Toyomarto village, but in general, hard work is necessary in order to be able to bring access to a piped water supply system for all residents.

12.3.2 Community Managed Water Supply System

This section explains the HIPPAM group as a voluntary association which might be categorized as a self-help group in providing water supply for members (Tao, 1999).

The common structure of the organization of HIPPAM in the research area, consisting of one chairman, one secretary, one treasurer and two technicians. The committee basically was selected by and from among the members of HIPPAM as a voluntary task to manage their water supply system.

The HIPPAM group has regular meetings, such as weekly or monthly meetings, to discuss their organization. It is interesting that the group likes to schedule water association meetings together with other community activities. For example, since the religious group has weekly meetings, along with the time for reciting the Quran, they will also set aside time to solve their water association problems. Through a kind of informal meeting, they are also able to develop their rules, norms, and sanctions to keep their voluntary association running well.

Herewith is the example of HIPPAM GOTONG ROYONG II in Sumberawan hamlet, with Mr. H. Toha as the chairman, Ms. Sunarti as the secretary, Mr. Sukar as the treasurer, and Mr. Sueb and Mr. Abdul Madjib as the technicians. This group has monthly meetings for all members. Until February 2009, there were quite a big number of members covering 137 households, and this group still has the ability to serve a maximum of 160 households, with the decision regarding the acceptance new members handled at board meetings. The group set up the cost at IDR 1,500,000 for the first physical pipeline installation of the new household water connection, and a fixed price of IDR 5,000 for monthly payment of water usage per household. This group has quite complete regulation as well as sanction to operate their community water management. The group has developed rules for one who wants to become a member as follows:

1. The resident must register with the committee;
2. Make full payment in accordance with the collective decision or the community agreement;
3. Comply with board rules.

If the candidate member agrees, he/she will sign a letter of statement signed by the head of village, the head of HIPPAM and the secretary. And then, if the candidate becomes a member, he/she will get a payment card. Next, every member must fulfill membership regulations regarding 10 issues, as follows:

1. It is forbidden to channel water into someone's house unless a permit is administered for a certain important purpose;
2. Water must be used sufficiently;
3. Faucets must be turned off when not in use;
4. Non-consumers are prohibited from using water for buildings or ceremonies without committee approval;
5. If no water flows, the user must report to the committee and should not change the stop valve or main pipeline by him/herself;
6. There should be no leaky pipe or faucet;
7. The consumer must pay for any individual house connection from the main pipe;
8. Dues must be paid each month as determined by the committee;
9. Compulsory support work and help to administrators are required;
10. It is obligatory for each member to keep harmony and engage in mutual assistance.

This group also stipulates a sanction system for members who do not comply with regulations. The type of sanctions could be (i) a temporary disconnection or (ii) a requirement to pay cash to the group at a minimum amount of IDR 5,000. And there are three steps of sanctions as follows:

1. Warning I (no water connection for a month);
2. Warning II (no water connection for 3 months);
3. Severance of water connection and loss of all costs.

12.3.3 Social Structure

12.3.3.1 Centrality in multiple memberships

The set of actors is denoted by $N = \{n_i, n_2, \ldots, n_g\}$ as a compilation of the respondents in each hamlet. There are three sizes of the g actors, consisting of (i) 100 respondents in Sumberawan hamlet, (ii) 64 respondents in Ngujung hamlet, and (iii) 70 respondents in RW 6 hamlet. And the set of events is denoted by $M = \{m_i, m_2, \ldots, m_h\}$ with h events consisting of 4 community associations, namely (i) Religious group, (ii) Cultural/Social group, (iii) Community Organization and (iv) Finance group. Thus this section illustrates the social structure of the respective respondents through multiple memberships in voluntary associations in the three different hamlets by analyzing the centrality of the affilliation network.

In the context of bipartite adjacency matrix, raw degree centrality for a respondent is simply the number of community groups in which they participate. This means the maximum raw degree centrality for a respondent in the three hamlets is 4, due to the total number of community groups. Table 12.2 illustrates 4 central respondents as the stars in the three hamlets. There are two stars in Ngujung hamlet, whereas the other hamlets have only one respondent exclusively. All of the four most central respondents have similar measures of degree centrality at $C'_D = 0.75$, whereby they are joint participants in three among four community groups. The result of degree centrality indicates that respondents in Ngujung hamlet have the highest level, followed by Sumberawan hamlet and RW6 hamlet. Thus, from the overall result, it may be concluded that the central respondent in Sumberawan hamlet plays a more important role compared to the other two hamlets, even when compared to Ngujung hamlet, since, particularly, in this hamlet, there are two respondents occupying the highest level. In other words, the result may indicate that in Sumberawan hamlet, power relations have a more significant meaning among respondents than among respondents elsewhere.

Table 12.2 Estimation result of centrality in multiple memberships.

	Sumberawan			Ngujung			RW 6		
	C'_D	C'_B	C'_C	C'_D	C'_B	C'_C	C'_D	C'_B	C'_C
Mean	0.300	0.0007	0.956	0.320	0	1	0.232	0.0041	0.858
Min	0	0	0.530	0	0	1	0	0	0.505
Max	0.750	0.0065	1	0.750	0	1	0.750	0.0708	0.981
Std. Dev.	0.163	0.0014	0.080	0.169	0	0	0.172	0.124	0.125
Variance	0.027	0	0.006	0.029	0	0	0.030	0.0002	0.016
Level of centrality (no. of respondent)									
0–0.333	71	89	0	42	0	0	58	52	0
0.334–0.666	28	0	3	20	0	0	11	0	6
0.667–1	1	0	86	2	0	58	1	0	46

In general, almost all respondents in the three hamlets have direct contact with every other respondent. Therefore, the betweenness centrality for each respondent in the three hamlets is almost zero. This indicates that there is no respondent occupying the position of mediatory between others. In this sense, it may be surmised that there is no bridging social capital between respondents in each hamlet. On one hand, it may be concluded that the respondents in each hamlet tend to form closed networks in which they tend to rely on their internal group. On the other hand, they might find difficulty when they have a problem beyond the group capacity to solve it, due to the lack of connections to groups external to their community. Moreover, it might not be easy to introduce 'new knowledge or innovation' once residents believe that their current circumstances are the most suitable for them.

Since closeness centrality is applicable only to defining the presence of the shortest path between the actors, in the case of an isolated actor whose position in the graph is disconnected, the actor becomes unreachable and the closeness centrality will be undefined. Therefore, it is necessary as a first step to exclude some respondents who engage in no joint participation in any kind of community group. Hence, it may be assumed that the influence of one respondent on another declines with the distance between them, whereby the minimum distance formulation of respondent closeness may give more effective contagion information to others. In this sense, it may be seen that transferring information among respondents in RW 6 will occur along the shortest path since each respondent is adjacent to all other respondents. Then it is followed by Sumberawan hamlet in the second position and it is closed by Ngujung hamlet. Next, it may also be seen that every respondent in Ngujung hamlet is maximally close to all other respondents, with equal influence in social interaction to each other. In Sumberawan hamlet, the average closeness of the respondents is 0.956, which means that almost every respondent has almost a path linking to all the others. Additionally, there is one respondent who has maximum closeness, in that he can reach all the other actors in a minimum number of steps. He is a driver 50 years old with a level of income on average between IDR 500,000 – 1,000,000 per month, who has been in the hamlet since he was born. Moreover, he has an educational background of high school, his household consists of 4 family members, and they are members of HIPPAM group. The average closeness centrality of respondents in RW 6 is slightly lower compared to the other two hamlets at 0.858. But, again in general the respondents have an almost equal number of minimum steps to reach all other respondents.

The structure of the social network of the residents in Sumberawan hamlet may be described as a combination of (i) one central actor occupying the position of formal and informal influential leader with possibly a high ability to influence villagers through minimum steps when relating to all others, (ii) very low occurrence of or almost no one in the position of mediator for a pair of residents, (iii) middle and high level of closeness to each other, and (iv) all respondents have ties to each other through their membership in at least one community group. Hence, an assumption may develop that the single choice for a piped clean water supply system from HIPPAM is a result of respondents' collective action driven by a central leader, since in general all respondents may have considerable equal influence on each other within the community.

The structure of the social network of the residents in Ngujung hamlet may be illustrated as a combination of (i) less power relation due to two different informal leaders with different choices of piped clean water supply systems, (ii) no one in the position of mediator since (iii) every respondent is adjacent to every other respondent within the internal community, and (iv) all respondents have ties to each other through their membership in at least one community group. And, in the case of RW 6, the social structure of the residents is a combination of (i) one informal leader who has equal distance of influence as the others, (ii) very low betweenness – but slightly higher than among the respondents in Sumberawan, (iii) middle and high level of closeness to each other, and (iv) moderate cohesion among respondents since some of them do not participate in any kind of community group. Then, it may be assumed that their choices regarding a piped clean water supply system are the result of individual action, whereby among residents there might be quite high connectedness and an equal ability to influence each other, but there is no major power relation enabling them to derive a common option. In other words, there is social interaction that may influence their decision but at the same time there is no central influential actor able to encourage them to behave homogenously.

If the central respondent is considered the influential actor with a high ability to influence others, then the central respondent may be categorized as one who has the role of informal leader due to that respondent's central position in the network. It is interesting that the central respondent in Sumberawan hamlet is very different from the others, particularly in terms of occupation. This central respondent is the head of Sumberawan hamlet and at the same time occupies the position of secretary of the HIPPAM association at the level of Singosari district. In this sense, the central respondent is the unified embodiment of both formal and informal leader. In the case of Ngujung hamlet, there are two stars, both of them female with occupations related to entrepreneurship, though on a very small, home-based business scale. It is interesting that they joined different piped clean water supply systems (PDAM, and HIPPAM). In this sense, they can be categorized as occupying the position of informal leaders through their wide contacts with others, but their influence may not be as strong as that of the leader in Sumberawan hamlet. Next, there is one central respondent in RW 6 with the occupation of farmer and a monthly household income of less than IDR 500,000. In this sense, it may be guessed that the role of the leader in RW 6 may not be as high as the leader in Sumberawan hamlet, but it might be higher than the leader in Ngujung hamlet. Thus, it may be guessed that the central actor in Sumberawan hamlet has the highest influence in his community compared to other stars. Hence, Sumberawan hamlet's central actor may be followed by the central actor in RW 6, with the two central actors in Ngujung hamlet last in order of community influence.

12.3.3.2 Social Networks

Using a Yes or No answer, the respondent was requested to answer: "Do people in the community generally trust one another in matters of lending and borrowing?" The terms *lending* and *borrowing* here refer to lending and borrowing for daily needs such as amounts of money for urgent needs – for hospital fees, for instance, for rice before harvest time, or just for simple things such as domestic tools.

Among the three hamlets, Table 12.3 illustrates that 90% of respondents in Sumberawan hamlet have trustworthiness among them about lending and borrowing. This was followed by 88% of the respondents in Ngujung hamlet and 87% of the respondents in RW 6. It may be argued that the high level of trust regarding lending and borrowing among communities tends to occur in the hamlets where the majority of households become members of HIPPAM. This is a form of social capital, the strong social ties that might foster in villagers the ability to mobilize local resources in order to pursue a common goal.

Table 12.3 Trust on lending and borrowing.

	Sumberawan		Ngujung		RW 6	
	Yes	No	Yes	No	Yes	No
HIPPAM	86	10	56	9	40	6
PDAM	-	-	1	0	16	2
Individal Well	-	-	1	0	-	-
Public Hydrant	-	-	3	0	-	-
Other	4	-	-	-	-	-
Total	100		64		70	

Following the previous question, the respondents were asked, "In the last three years, has the level of trust has improved, worsened, or stayed the same?" Below, Table 12.4 illustrates that only 14% of respondents in Sumberawan hamlet, around 16% respondents in Ngujung hamlet, as well as in RW 6 reveal that their trust has been getting worse within the last three years. In other words, it might be assumed that trust among villagers is at the same level or has even improved in the last three years for all three hamlets. It may be argued that regarding social ties formed between households within the same community, membership in organizations such us HIPPAM preserve their ability to sustain trustworthiness.

Table 12.4 Level of trust in the last three years.

Hamlet	Sumberawan			Ngujung			RW 6		
	I	S	W	I	S	W	I	S	W
HIPPAM	28	55	13	27	13	6	17	38	10
PDAM	-	-	-	12	2	4	0	1	0
Individual Well	-	-	-	-	-	-	1	0	0
Public Hydrant	-	-	-	-	-	-	2	0	1
Other	3	0	1	-	-	-	-	-	-
Total		100			64			70	

In the questionnaire survey, there are 13 questions to describe respondents' opinions about their living environment, on a scale of 1 to 5, where 5 means very much and 1 means very little, as follows:

A) I feel proud of this area/community
B) I think nature and scenery in this area/community are great
C) I think the availability of foods in this area/ community is great
D) It is important to participate in local events
E) It is important to collaborate to make this area/community better
F) It is important to join in educational activities for the growth of local children
G) It is important to listen to others' problems
H) It is important to maintain a daily relationship with neighbors
I) It is important to take care of a single or elderly person
J) It is important to maintain regional graveyards
K) It is important to manage relationships with your relatives in the area/community
L) People in this area/community are important to me
M) I want to continue living in this area/community

Table 12.5 Perspective on the living environment.

	A	B	C	D	E	F	G	H	I	J	K	L	M
Sumberawan													
HIPPAM	5	3	5	5	5	5	5	5	5	5	5	5	5
Other	5	3	5	5	5	5	5	5	4	5	5	5	5
Ngujung													
HIPPAM	5	4	5	5	5	5	5	5	5	5	4	5	5
PDAM	5	5	3	5	5	4	5	5	5	5	5	5	5
RW 6													
HIPPAM	5	4	4	5	5	5	5	4	4	5	5	5	5
PDAM	5	4	5	5	5	5	5	5	5	5	5	5	5
Individual Well	4	3	4	5	5	5	5	5	4	5	5	5	4
Public Hydrant	na	4	5	5	na	5	5	na	4	5	5	na	na

Firstly (as can be seen in Table 12.5), respondents in Sumberawan hamlet indicate that HIPPAM members have slightly higher respect for their living environment than others who utilize water from the river, at an average 4.85 compared to 4.77. Secondly, respondents in Ngujung hamlet illustrate that HIPPAM members also have slightly higher respect for their living environment than PDAM customers, at the same average as for the two groups in Sumberawan hamlet. Thirdly, respondents in RW 6 descirbe PDAM customers as showing the highest respect for their living environment at an average of 4.92, more than the second highest, the HIPPAM members who averaged 4.69, and the third highest, the households utilizing individual wells, who averaged 4.54. There were 5 items left unanswered by households utilizing water from public hydrants, so that this issue cannot be examined due to imperfect reflection toward the whole perspective of this group of households.

Furthermore, villagers who acquire access to a piped water supply system, whether it comes from HIPPAM or PDAM, in general have a higher respect for their living environment than villagers who have no access to piped water, instead utilizing water from rivers, individual wells, and public hydrants. This indicates that villagers who have the ability to mobilize or utilize local resources for their betterment have a tendency to respect their living environment more highly than those without this ability.

Sumberawan hamlet, where villagers choose to acquire water from HIPPAM and the river, and Ngujung hamlet, where villagers choose to get access to piped water either from HIPPAM or PDAM, seem to have similar circumstances in that villagers generally have a high respect for their living environment. In general, this indicates that all villagers in Ngujung hamlet have access to a piped water supply system, whether they tend to be a member of the HIPPAM group or a customer of PDAM. On one hand, the villagers in Sumberawan hamlet who belong to the HIPPAM group may have access to a piped water supply system that is maintained by the community itself. Other villagers' habit of acquiring water from the river may reflect their traditional way of utilizing water as a local resource. This might answer why differences in the method of water utilization by the villagers in the same hamlet may not indicate differences in their level of respect for their living environment. Meanwhile, villagers in RW 6 have four types of access to water. Some villagers get access to a piped clean water supply system through HIPPAM or PDAM, while others acquire water from a non-piped source, an individual well or public hydrant. In this hamlet, villagers show quite different levels of respect for their living environment. Villagers with access to the piped water supply system in general have slightly higher respect than those without access. In this sense, it may be argued that the ability to mobilize local or extra-resources might make villagers have slightly higher respect for their living environment than villagers without this ability.

In other words, it may be argued that all villagers in general have high respect for their living environment. Since they consider their life as part of nature, they always try to live in harmony with nature. In this sense, there needs to be more detailed scrutiny of how social ties as the personification of local wisdom may maintain villagers and their living environment harmoniously.

12.4 CONCLUSION

There are three types of social structure of the community in the research area. The first community is best described as a combination of (i) the existence of power relations driven by a formal and informal leader and (ii) a strong social tie within the community which enables members to establish and maintain their community-based water management. The second is a community where (i) the existence of power relations is inadequate to drive community members to reach a common goal, but (ii) they have moderately strong social ties within the community, with a lack of bridging social capital. This situation may cause community members' current collective action of managing their water supply system to be difficult to sustain in the long term. The last community could be described as an area with characteristics of both communities. In this community, there is a power relation but it is not as strong as in the first community, as the central figure is simply occupying the position of informal leader. Moreover, in general there are moderately high social ties within the community that might enable the community to reach their common goal driven by demographic characteristics such as educational background and quite wide contact with actors outside the community. Also important is the fact that this community is able to 'formalize' its community-based water supply system. In this sense, their social ties might bring them the ability to maintain their community-based water supply system in a way different from that of the first community.

Thus, it may be concluded that investigating respondents' multiple memberships in community groups may provide evidence of a linkage between collective action theories and the social capital approach, that the choice to join a community-based water supply system may reflect individual choice as well as collective action. But it may be argued that the motivation for collective action is derived from the power relation initiated by a central respondent.

Joint participation in a kind of community-managed water supply system such as HIPPAM might strengthen social ties among members, helping them to form and maintain trustworthiness among themselves at a steady, high level. Through their ability to establish and maintain HIPPAM to pursue the common goal of having access to a piped water supply system also boosts the positiveness of social capital among villagers, enabling them to maintain trustworthiness and respect toward their living environment.

By examining the social structure of the community though three indices of centrality, this research focuses upon social ties created from emotional nearness through multiple memberships in the community group. However, this research does not fully examine the institution of the community-managed water supply system. Needless to say, it is necessary to investigate the institutional identity of voluntary associations of community-managed water supply systems to be able to propose appropriate water governance for more sustained collective action on managed water supply systems. These topics remain for future research.

12.5 ACKNOWLEDGMENT

Grateful acknowledgement is made to the Directorate General of Indonesian Higher Education, Department of National Education, Indonesia, the sponsoring institution for the 1st author, a lecturer at the University of Brawijaya, to pursue a doctoral degree at Kyoto University from 2008 to 2011.

12.6 REFERENCE

Aoki M. (2009). Community norms and embeddedness: a game-theoretic approach. In: M. Aoki. and H. Yujiro, (Eds), *Communities and Markets in Economic Development*, Oxford University Press, New York.

Bird J., Arriens W. L. and Custodio D. V. (2008). Water rights and water allocation: issues and challenges for the Asian region. *Network of Asian River Basin Organizations (NARBO),* Asian Development Bank (ADB), Philippines.

Brandes U. and Pich C. (2006). *Centrality Estimation in Large Networks*, Department of Computer and Information Science, University of Konstanz, Germany.

Candi Renggo Village Basic Data (2006-2008). Singosari District, Malang Regency.

Christian G., Deepa N., Nyhan J. V., Woolcock M. (2004). *World Bank Working Paper No. 18: Measuring Social Capital An Integrated Questionnaire*. The World Bank, Washington D.C.

Everett M. and Borgatti S. P. (2005). Extending centrality. In: P. J. Carrington, J. Scott and S. Wasserman (eds.), *Models and Methods in Social Network Analysis*, Cambridge University Press, New York.

Faust K. (1997). Centrality in affiliation networks. *Social Networks,* **19**, 157–191.

Freeman L. C. (1977). A set of measures of centrality based on betweenness. *Sociometry,* **40**, 35–41.

Malang Regency, *Candi Renggo Village Basic Data, 2006–2008,* Singosari District, Malang Regency.

Malang Regency, *Toyomarto Village Basic Data, 2006–2008,* Singosari District, Malang Regency..

PDAM Malang Regency (2008). *Data Base of Clean Water in Malang Regency*, Malang Regency.

Putnam R. D. (2000). *Bowling Alone: The Collapse and Revival of American Community*, Simon and Schuster, New York.

Sabidussi G. (1966). The centrality index of a graph. *Psychometrika,* **31**(4), 581–603.

Statistic Centre Board Indonesia (2005), Badan Pusat Statistik, Jakarta.

Tao M. (1999). *Voluntary Association Management*, Tokyo.

Toyomarto Village Basic Data (2006-2008). Singosari District, Malang Regencey.

Wasserman S. and Faust K. (1994). *Social Network Analysis: Methods and Applications*, Cambridge University Press, New York.

WHO (2010). *Access to safe drinking water improving; sanitation needs greater efforts*, WHO, Geneva, New York.

WHO and UNICEF (2006). *Meeting the MDG Drinking Water and Sanitation Target: The Urban and Rural Challenge of the Decade*. WHO and UNICEF, Switzerland.

13 A cognitive analysis of residents with regard to community-based flood management

Surjono[1], Misbahib Haraha[2], Jenny Ernawati[3]

[1] Department of Urban & Regional Planning, University of Brawijaya, Jl MT Haryono 167 Malang, Indonesia. E-mail: surjono@ub.ac.id

[2] Department of Civil Engineering, University of Brawijaya, Jl MT Haryono 167 Malang, Indonesia

[3] Department of Architecture, University of Brawijaya, Jl MT Haryono 167 Malang, Indonesia

Abstract

Issues of water and social capital are associated not only with the provision of clean water but also with natural disaster mitigation, since rainfalls often create floods in most Indonesian cities. This paper investigates the social capital of three shareholders with regard to flood hazard mitigation in Indonesia: local residents, local government, and school communities. The capacity index of flood hazard preparedness was evaluated using 5 variables, based on UNESCO-ISDR and LIPI (Indonesian Science Institute) guidance: knowledge and attitude (KA), policy statement (PS), emergency planning (EP), warning system (WS) and resource mobilization capacity (RMC). As a case study, we looked at the experience of the city of Pasuruan, East Java Province. Pasuruan is a representative of medium-size cities located in coastal areas and threatened by floods every year. The assessment found that Pasuruan was inadequately prepared, and this was caused by capacity weaknesses of the local residents, school communities, and local government. We recommend that empowering and strengthening be focused on the weakest sub-variables at the local resident level, after which there should be a strengthening of capacity and connectivity among the three shareholders.

Keywords: social capacity, preparedness index, flood mitigation

13.1 INTRODUCTION

Jones and Clark (2013) argued that social capital's impacts on the risk management of natural disaster are under-researched. Global climate change has contributed to the awareness of disaster prevention and mitigation research, particularly for areas prone to climate change-related hazards, such as coastal areas, which are sensitive to the rise of sea level. Current research interest, however, has been focused on physical mitigation technology and less on the social aspects of the mitigation system. Current interest is generated by government policies, while the acceptability of the policies is significantly determined by community social capital.

There is a significant linkage between disaster mitigation management and social capital, particularly when disasters are partly caused by human activities. Human activities are not the causes of disaster if they are conducted in an environmentally friendly manner. Development practices, however, have become the causes of hazards and created risk for people. In many cases poor people and individuals with lower levels of social capital experience higher risk (Briceno, 2008). All parameters of social capital, which is comprised of social trust, institutional trust, social norms, and social networks (Jones *et al.* 2010), should be strengthened, with institutional trust being the most important in terms of hazard response (Jones *et al.* 2012). Trust among institutions in local government, private organizations and communities, therefore, must be developed in order for shareholders within the city to strengthen the city's capacity for disaster mitigation.

Among institutions, awareness of potential hazards is crucial in generating stakeholders who are responsive to mitigating disasters. Disaster management in Indonesia is coordinated by a non-structured institution, the National Coordination Body for Disaster Mitigation; however, this institution is often unable to handle the complexity and level of disaster in Indonesia. This situation is worsened by the indication that bureaucracies and communities have little awareness or basic understanding of disaster management. Therefore, increasing the roles of bureaucracy and community is significant, since community may behave as an autonomous actor with its own interests, preferences, resources, and capabilities (Patterson *et al.* 2010).

Social capital in Indonesia tends to be bonding rather than bridging social capital. Wolf *et al.* (2010) and Paterson *et al.* (2010) stated that social cohesion or bonding social capital in a community may create complex relationships and increase rather than reduce vulnerability to potential disasters. Bonding social capital leads to a behavior that may be irresponsive to incoming hazards, and as researchers have argued, people's behavior can be positively influenced by improving social networks (Schelong, 2007; Mathbor, 2007). It is important to take into account human behavior when exploring ways to improve disaster management strategic policies. As proposed by many developed governments, such as Queensland Government (2010), preparedness of the community as an element of the disaster management strategic policy framework will strongly influence the degree of impact of a disaster. In the case of Indonesia, a large part of the Indonesian community lacks knowledge of hazard and disaster prevention (ISDR, 2007). So, it is important to measure the preparedness capacity within the society in relation to the disaster management policy framework in Indonesia. In the context of flood disaster management, preparedness may begin from the observation of how the community prepared for prevention, the response/intervention, and the recovery stage. This research aims to evaluate the capacity level (index) of preparedness with regard to flood hazard in urban settlements. The analysis evaluated five variables of the community's capacity:1) knowledge and attitude, 2) policy statement, 3) emergency planning, 4) warning system, and 5) resource mobility.

13.2 METHODS

The capacity index of preparedness in this research was constructed from five variables:

- Knowledge and attitude (KA) comprise four variables: 1) knowledge about floods; 2) knowledge about environmental vulnerability; 3) knowledge about physical buildings' vulnerability and significant facilities for disaster emergency situations; and 4) attitude towards flood risk.

- Policy statement (PS) consists of three variables: 1) Types of preparedness to anticipate flood hazards; 2) relevant regulations; and 3) relevant guidelines.
- Emergency Planning (EP), consists of eight variables: 1) disaster management organization; 2) evacuation planning; 3) disaster camp and fixed procedure; 4) initial help plan, evacuation, safety and security during disaster; 5) basic need fulfillment plan; 6) refuge, clean water and sanitation, health and disaster information; 6) evacuation equipment; 7) important facilities for emergency situations; 8) exercise and evacuation simulation.
- Resource Mobilization Capacity (RMC) consists of six variables: 1) institutional arrangement and command system; 2) human resources; 3) technical assistance and provision of material for flood preparedness; 4) financial mobilization; 5) coordination and communication among stakeholders; 6) monitoring and evaluation of disaster preparedness activities.
- Warning system (WS) consists of three variables: 1) traditional warning system; 2) technologically based warning system; 3) exercise and simulation.

Calculation of the index used the following equation:

$$\text{Index} = \frac{\text{real total score}}{\text{maximum score}} \times 100$$

As proposed by Britton and Dynes & Drabek in Gurtner *et al.* (2011), there are five basic social units in social research assessment paradigms, i.e.: 1) individual, 2) family, 3) organization, 4) community, and 5) society and/or international systems. In this research we looked at the capacity of family, community and society, which was represented by three indices: index of resident (households/neighbourhood) (R) (35% weight) + index of government (G) (35% weight) + index of school community (SC) (30% weight) (LIPI-UNESCO-ISDR, 2006).

13.2.1 Index of Residents (Individuals & Households) (R)

The index of residents was evaluated from 97 samples (n) from 449 households (N) obtained from Slove's formula, i.e. $n=\frac{N}{1+Ne^2}$. The R index was calculated with the following weight of indices: (0.45*KAR index) + (0,35*EPR index) + (0,15*RMCR index) + (0,05*WSR index).

13.2.2 Index of Government (G)

The index of government consists of three components, i.e. local government (G1); technical institutions/bodies at the local government (G2) represented by 15 heads of technical institutions; and district government (G3), two staffs in each district. Each component was evaluated by 53 parameters of preparedness.

- G1 index= (0.25*PSG1 index) + (0.34*EPG1 index) + (0.28*RMCG1 index) + (0.13*WSG1 index)
- G2 index= (0.74*KAG2 index) + (0.07*EPG2 index) + (0.11*RMC G2 index) + (0.07*WSG2 index)
- G3 index= (0.35*PSG3 index) + (0.25*EPG3 index) + (0.35*RMCG3 index) + (0.05*WSG3 index)

- KAG index=KAG2 index
- PSG index= (0.65*PSG1 index) + (0.35*PSG3 index)
- EPG index= (0.72*EPG1 index) + (0.08*EPG2 index) + (0.20*EPG3 index)
- RMCG index= (0.60*RMCG1 index) + (0.12*RMC G2index) + (0.28*RMC G3 index)
- WSG index= (0.70*WSG1 index) + (0.20*WSG2 index) + (0.10*WSG3 index)
- G index= (0.20*KAG index) + (0.20*PSG index) + (0.25*EPG index) + (0.25*RMCG index) + (0.10*WSG index)

13.2.3 Index of School Community (SC)

The school community index was grouped into three components, i.e., school index (SC1); teacher index (SC2); and student index (SC3). Each component was evaluated according to 34 parameters of preparedness.

- SC1 index= (0.29*PSSC1 index) + (0.41*EPSC1 index) + (0.18*RMC SC1 index) + (0.12*WS SC1 index)
- SC2 index= (0.71*KASC2 index) + (0.17*EP SC2 index) + (0.07*RMC SC2 index) + (0.05*WSSC2 index)
- SC3 index= (0.83*KA SC3 index) + (0.08*EP SC3 index) + (0.04*RMC SC3 index) + (0.04*WSSC3 index)
- KASC index= (0.6*KASC2 index) + (0.4*KASC3 index)
- PSSC index= PSSC1 index
- EPSC index= (0.61*EPSC1 index) + (0.30*EP SC2 index) + (0.09*EP SC3 index)
- RMCSC index= (0.60*RMC SC1 index) + (0.30*RMC SC2 index) + (0.10*RMC SC3 index)
- WSSC index= (0.57*WS SC1 index) + (0.29*WSSC2 index) + (0.14*WSSC3 index)
- SC index= (0.50*KASC index) + (0.10*PSSC index) + (0.23*EPSC index) + (0.10*RMCSC index) + (0.07*WSSC index)

Preparedness of the society was classified into five scales (Table 13.1):

Table 13.1 Scale of preparedness.

No	Index score	Scale of preparedness
1	80–100	Very prepared
2	65–79	Prepared
3	55–64	Nearly prepared
4	40–54	Less prepared
5	0–39	Not prepared

Source: LIPI-UNESCO-ISDR (2006)

13.3 RESULTS

The capacity of Pasuruan's society to prepare for flood risk and disaster can be seen from the results of the preparedness index that consists of three indices: residents (individuals and/or households) index (R), government index (G), and school community index (SC).

13.3.1 Capacity of Residents (Individuals/Households)

The total index of the preparedness of individuals and households in Pasuruan City to anticipate floods, with regard to five variables, was 12.11. Based on this scale, the capacity of individuals/households can be termed 'not prepared'. The capacity of preparedness according to each variable is shown in Table 13.2.

Table 13.2 Scale of preparedness of households (R).

No	Variables	Scale of preparedness
1	KAR index	20.70
2	EPR index	7.08
3	RMCR index	0.58
4	WSR index	4.55
Total R index		12.11

Pasuruan city is comprised of three districts, and it has four villages that are impacted by 3 to 7 floods per year. The impact of such floods covers about 360 ha of settlement in Pasuruan. Table 13.2 indicates that frequent accidents due to flooding have led to an increase in knowledge about the causes of floods. However, the community at the household level did not have adequate capacity to mobilize resources. The community is becoming accustomed to seeing floods as a routine environmental feature. This perception actually increases the residents' vulnerability, particularly those who live on the riverbanks of the three rivers in Pasuruan. Table 13.2 also indicates that warning systems and emergency planning within the community have been poor. The warning system was based only on weather changes and communication among neighbors, without any emergency planning related to imminent disasters/big floods.

13.3.2 Capacity of the Local Government

Quantitative analysis to measure the capacity of the local government to prepare for any risk of flood showed that the scale of the local government of Pasuruan was 'nearly prepared'. Indices obtained from the survey indicate that the least prepared variable was the warning system (WSG index). The score for each variable is shown in Table 13.3.

Table 13.3 Scale of preparedness of local government (G).

No	Variables	Scale of preparedness
1	KAG index	62.30
2	PSG index	64.37
3	EPG index	62.12
4	RMCG index	66.68
5	WSG index	37.27
Total G index		56.96

Table 13.3 indicates that the lowest score was for the warning system variable. This shows that the Pasuruan government did not prepare a warning system for floods, a problem stemming from the issue of authority over river management under current governance. The three rivers causing the floods are the responsibility of the provincial government, with local governments lacking the willingness or initiative to monitor and maintain them. The mechanisms that have not worked well are networks and institutional trust among governments at the local level.

13.3.3 Capacity of School Community

The total index of school community was equal to 50.47. This value is categorized as 'less prepared'. Among the five variables, resource mobilization capacity (RMCSC), as also occurred in households' capacity, was the lowest, i.e., 10.59, while the highest was the knowledge and attitude (KASC) variable (see Table 13.4).

Table 13.4 Scale of preparedness of school community (SC).

No	Variables	Scale of preparedness
1	KASC index	63.42
2	PSSC index	50.00
3	EPSC index	47.69
4	RMCSC index	10.95
5	WSSC index	25.58
Total SC index		50.47

Floods almost never reach most schools in Pasuruan, due to their location. Therefore, school communities never prepare for flood risk. Also, schools never conduct emergency evacuation simulation practices, or train teachers or students to mitigate floods. So, it is understandable that the RMCSC index was low.

13.3.4 Total Capacity of Community

The aggregate index of the capacity to anticipate floods in Pasuruan was 46.60, which fell under the 'less prepared' capacity. Among the three shareholders of the community, i.e., the community at the residential level, the local government, and the school community, the most prepared shareholder was the local government, while the least prepared was residents' capacity (Figure 13.1).

Figure 13.1 shows that local government had the highest score in total and was very dominant in the five variables. Conversely, capacity at the residential level was the weakest. This proved that social networks and norms failed to create balanced capacity among the three shareholders of the community. The concept of UNESCO, indicated from the distribution of weight in the scoring system, promotes a balanced capacity among residents (35%), local government (35%) and school communities (30%). This means that the residential/neighbourhood, school community and local government should contribute equally to flood mitigation. Capacity at the residential level, at only one fifth of local government's capacity, shows that linkage among the three social units has not worked well.

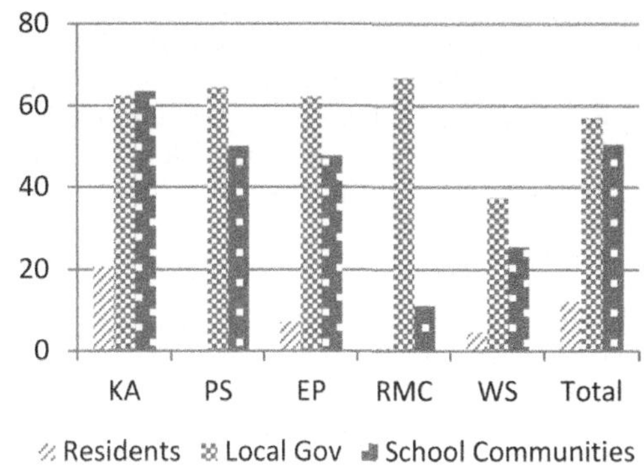

Figure 13.1 Total index of preparedness.

13.4 CONCLUSIONS

To some extent, the index of preparedness in flood mitigation management represents social capital within the community because it shows the linkage of agents and norms. As argued by Murphy (2007), disaster management at the local level needs interdependency among social units although they are separate aspects of management. Conditions in Pasuruan reflect the actual situation in Indonesia, where the community is too dependent on the government. Large sections of the society, represented by the index of residents (R), have lacked understanding of the risks posed by floods and the methods of protecting against them. There were significant gaps between the residents and the government. This showed that development at the local level has not created bridging social capital within the community. The knowledge that has been gained by the government and school community has not been disseminated to local residents. Such knowledge affects how a community manages emergency planning, warning systems, and the mobilization of resources.

Future research agenda in Indonesia should include seeking the missing link and strengthening the link between social units, so that development and improvement in one social unit will push other units to improve and create balanced capacities. It is hoped that residents will achieve substantial capacity in preparedness. The community at the residential level should increase its capacity to be equal to the other two components of the society (school community and local government). Bonding social capital in the community, which is strong in Indonesian culture, should be redirected to form bridging social capital by, e.g., improving the command system, forming trained voluntary personnel, and strengthening trust between the community and government.

13.5 REFERENCES

Briceno S. (2008). Disaster risk reduction as a contribution to inclusive education. *The 48th International Conference of Education (ICE)*, ISDR, Geneva.

Gurtner Y., Cottrell A. and King D. (2011). *PRE & RAPID: Community Impact Assessment for Disaster Recovery*. James Cook University, Brisbane.

ISDR (2007). Towards a culture of prevention: Disaster risk reduction begins at school. UNESCO, Geneva.

Jones N. and Clark J. R. (2013). Social capital and climate change mitigation in coastal areas: A review of current debates and identification of future research direction. *Ocean & Coastal Management*, **80**, 12–19.

Jones N., Clark J. and Tripidaki G. (2012). Social risk assessment and social capital: A significant parameter for the formation of climate change policies. *The Social Science Journal*, **49**(1), 33–41.

Jones N., Evangelinos K., Halvadaki C. P., Iosifides T. and Sophoulis C. M. (2010, July). Social factors influencing perceptions and willingness to pay for market-based policy among solid waste management. *Resources, Conservation & Recycling*, **54**(9), 553–540.

LIPI-UNESCO-ISDR (2006). Kajian Kesiapsiagaan Masyarakat Dalam Mengantisipasi Bencana Gempa Bumi dan Tsunami, Jakarta.

Mathbor G. M. (2007). Enhancement of community preparedness for natural disasters. *International Social Work*, **50**(3), 357–369.

Murphy B. L. (2007, January 11). Locating social capital in resilient community-level emergency management. *Natural Hazards*, 297–315.

Patterson O., Weil F. and Patel K. (2010). The Role of Community in Disaster Response: Comceptual Models. *Population Resource Policy Review*, **29**, 127–141.

Queensland Government (2010). *Disaster Management Strategic Policy Framework. Emergency Management Queensland, Department of Community Safety*, The State of Queensland, Brisbane.

Schelong A. (2007). Increasing Social Capital for Disaster Response through Social Networking Services (SNS) in Japanese Local Governments. National Science Foundation, National Center for Digital Government, NCDG, Harvard.

Wolf J., Adger W. N., Lorenzoni I., Abrahamson V. and Raine R. (2010, February). Social capital, individual responses to heat waves and climate change adaptation: an empirical study of two UK Cities. *Global Environmental Change*, **20**(1), 44–52.

14 Climate change adaptive planning and conflict resolution strategy of Krabyakan spring utilization

Mustika Anggraeni

Urban and Regional Planning, Brawijaya University, 167th MT. Haryono Street, Malang, Indonesia.
E-mail: m.anggraeni@ub.ac.id

Abstract

The objectives of this study are to identify the impacts of climate change on water resources availability and the influence of climate change on water demand and food security, to examine the conflict that occurred between several beneficiaries of water resources in the study area, and to discuss the formulation of strategies for conflict resolution concerning water resource utilization and adaptation to climate change. The methods used for data processing were a regression model with the T-Test and the Chi Square Test. Secondly, qualitative description methods were used to explain the conflict between users and stakeholders of the Krabyakan Spring water. To analyze the conflict and to illustrate the causes and effects of conflict dynamics, a conflict matrix was used. The formulation of strategies was accomplished through a literature review and the implementation of chance analysis based on the characteristics of the study area and of the conflicts that have occurred. Based on the results of the study, it can be seen that climate change impacts water resources availability, significantly correlated, and that climate change has an influence on water demand, food security, and the conflict that occurred between several beneficiaries of water resources in the study area. Strategies for conflict resolution of water resource utilization and adaptation to climate change formulated from this study are the following: managing water distribution; strengthening the bargaining position of the beneficiaries; maintaining sensitivity to conflict dynamics in climate change policies and programming for Krabyakan Spring; drawing upon traditional adaptation and coping mechanisms to determine the policy ultimately settled upon by the government; and Integrating climate change in development policy and strategies.

Keywords: adaptive planning, conflict resolution, climate change, water utilization.

14.1 INTRODUCTION

Water is one of the basic needs and is classified as merit goods where its presence is a necessity both in dry and rainy seasons. In some places, both urban and rural, clean water is a problem not easily solved, complicated by the availability of very limited water resources and the costs and technology required to process water before it is used by various people for various needs.

In urban and rural areas in East Java Province today, about 60% of people still live in conditions where it is very difficult to get clean water. Based on observations of lithology and soil texture in the region of East Java, ground water and soil water potential, as well as the potential for surface water as base water, is very low and unevenly distributed, resulting in many urban and rural areas becoming vulnerable to water supply problems (Anonymous, 2003). Krabyakan Spring is one of the water resources supporting the inhabitants of Malang and Pasuruan Region. Supply and demand problems related to water resources have occurred with population growth and the rapid rise in demand for water for use by households, industry, agriculture farms and embankment. To the contrary, the supply of water resources, especially in the Welang River shed in the Spring Krabyakan Drainage Area, does not meet the needs of human development. This issue has been developing for several years, with farmers complaining of water shortages year after year.

Society is already facing severe problems with water resources, in the form of excessive demand and insufficient supply. The governments of Malang and Sidoarjo have a Memorandum of understanding to share the use of water resources from Krabyakan Spring. This shared water resource is the trigger that generates unsolved conflict between beneficiaries of the water resources, Malang, Pasuruan and Sidoarjo. But this phenomenon does not yet include climate change symptoms that can worsen the problem.

Strong scientific evidence indicates that the average temperature of the earth's surface is increasing due to greenhouse gas emissions. The average global temperature has increased by about 0.6°C since the late 19th century (Smith and Vivekananda, 2007). The latest IPCC scenarios project temperature rises of 1.4–5.8°C, and sea level rises of 9–99 cm by 2100 (Houghton *et al.* 2001). Warming and precipitation are expected to vary considerably from region to region as the greatest increase in temperature is expected in the Northern Hemisphere.

Changes in climate averages, coupled with change in the frequency and intensity of extreme weather, are likely to have a major impact on natural and human systems. The effect on the world's poor in tropical and subtropical areas is likely to be disproportionately large, in particular since their potential to adapt to such changes is low (Smith, 2003). Many of the world's countries already struggle under existing water stress from pressures such as irrigation demands, industrial pollution and water borne sewerage. These pressures will be significantly exacerbated by climate change, which for many regions will result in reduced rainfall and increasing temperatures, further reducing the availability of water for drinking, household use, agriculture and industry. As these competing demands intensify under climate change, effective governance for balancing water demands will become essential, particularly in the face of strong pressures to prioritize industrial uses over other uses such as drinking supplies.

Agricultural use is the largest consumer of water around the world, accounting for 67% of all withdrawals and 79% of all water consumed (FAO, 2003). This is true as well of the study area, where the largest proportion of water demand is for agriculture needs. The lack of water resources, especially in the dry months, tends to be a threat to society, specifically for farmers, because it is related to food safety. The scarcity of water resources has become the trigger for competition and conflict. In the study area, conflict that has already occurred needs to be resolved in a prudent way because water is a very sensitive matter for society.

There is very little academic research on the links between climate change and conflict. Of the few studies that are available, the findings are mixed. Most authors argue that whilst climate change may well result in increased conflict, this path is not certain. The most commonly discussed scenarios are that climate change causes resource scarcity, which prompts violent conflict, or that resource scarcity results in migration which leads to conflict in the receiving area. In 2007, International Alert published a report – A Climate of Conflict – that both established the links between climate change and conflict and identified the outlines of a policy response (Smith and Vivekananda, 2007). The present report takes that analysis as its starting point and will not repeat the evidence and arguments that show that the knock-on consequences of climate change include a serious risk of increased violent conflict. Although there were doubts about the direct link between climate change and conflict, this study attempts to find symptoms. This study is needed because conflict is always inherent in life, especially in Indonesia. Hence we need adaptive planning that is conflict-sensitive.

The objective of this study is to identify climate change impacts on water resources availability, its influence on water demand and food security, and its connection with conflict that has occurred among several beneficiaries of water resources in the study area. The next goal is the formulation of strategies for conflict resolution of water resource utilization and adaptation to climate change.

14.2 METHODS

The location of the study was the trans-boundary between Regency of Malang, Pasuruan and Sidoarjo. The water resource is located in Sumber Ngepoh, Lawang, Malang Regency. This study was focused mainly on water utilization from Krabyakan Spring, climate change effect on it, and conflict among users to whom the water was distributed under three different management authorities. Field observation and secondary surveys were used as data acquisition methods. Rainfall and river debit data for 13 years were obtained from Purwodadi Climatic Station. Data collection methods to understand the conflict between stakeholders were interviews with each stakeholder about his or her interest and fears in the context of Krabyakan Spring water utilization.

The methods used for data processing were statistic methods to find the connection between climatic changes and surface water debit. Statistic methods used were a regression model with the T-Test and the Chi Square Test. Secondly, qualitative description methods were used to explain the conflict between users and stakeholders of the Krabyakan Spring water. For a more detailed analysis of the conflict, a conflict matrix was used, to illustrate the causes and effects of conflict dynamics. The formulation of conflict resolution of water resource utilization and adaptation to climate change strategies used a literature review and implementation chance analysis based on the characteristics of the study area and the characteristics of the conflicts that occurred.

14.3 RESULT AND DISCUSSION

The Krabyakan Spring, located in Sumber Ngepoh Village, has 7 sources that flow into the river channel. Measurement results obtained 2 variations in both rivers' discharge (cumulative inflow of the 7 sources): 745.736 liters/second and 378.470 liters/sec. The cumulative debit of the two rivers are 1124.2064 liters/sec. Utilization of water resources at this time is by everyday people who waged a kolaman used for bathing, washing Ngepoh Village community. It is also used to irrigate rice fields around the village and along the watershed Ngepoh Welang River Basin, which is the source Krabyakan used to meet irrigation needs in Pasuruan. (Figure 14.1. Service Area of Sumber Ngepoh Spring to Welang River). Rainfall and debit flow of the surface water of Welang River are recorded in the following table.

Figure 14.1 Service area of Krabyakan Spring to Welang Watershed.

The pattern of climate variability and average river debit can be seen in Figure 14.2 and Figure 14.3.

14.3.1 Climate Change Impact on Water Resources

The first objective of the research is to identify climate change impacts on water resources availability, and its influence on water demand and food security. First of all, based on the rainfall and the debit average in Table 14.1, we need to know how significant the correlation is between the two parameters. A regression analysis shows that the correlation of these two parameters is significant. This conclusion results from the chi square Test (Table 14.2). From Table 14.2, it can be concluded that the regression model is $y = 2{,}2+0{,}001x$. With α 5 %, the T-test also shows significantly that rainfall pattern change is correlated with the average debit of Welang River.

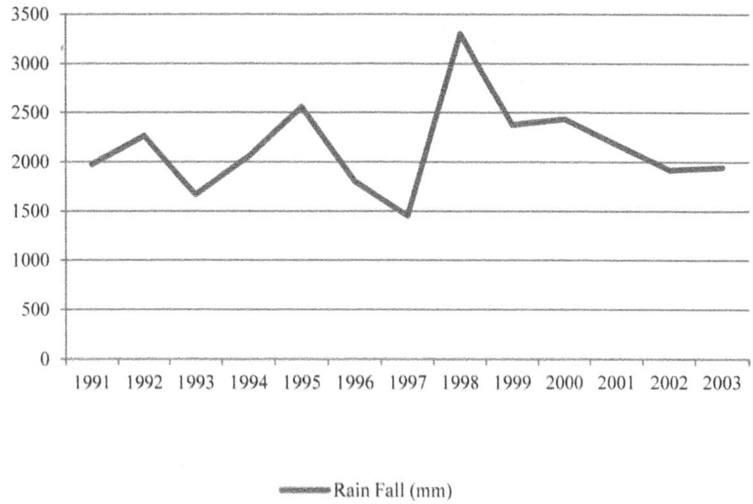

Figure 14.2 Pattern of rainfall variability.

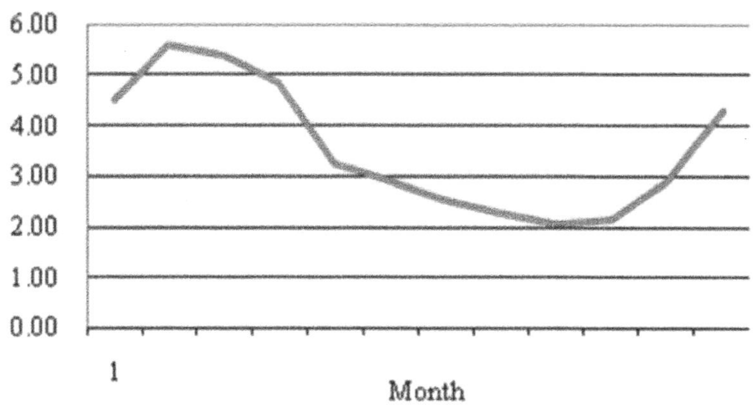

Figure 14.3 Average debit in Welang Watershed (m3/sec).

The use of water resources in the drainage area of Welang is mostly to meet agriculture demand, domestic needs, embankment, and industrial demand. The total demand for water increased along with population growth. The supply-demand balance of water resources in the drainage area of Krabyakan Spring is provided in Figure 14.5. This figure shows that there is an imbalance in supply and demand, in 90% availability of water, the demand exceeding available water supply, especially in June, July, and August. The prevalent influence on the supply-demand balance of Welang Watershed is irrigation water. As we know from Figure 14.4, the major land use for the study area is agriculture. The farming patterns and farming intensity that influenced the balance model are the following:

- Wet season: paddy (100 %)
- Dry Season I: paddy (50 %) secondary crops (50 %)
- Dry Season II: secondary crops (75 %).

Table 14.1 Annual rainfall (mm) and average Welang River debit (m^3/s).

Year	Annual Rain Fall	Average River Debit
1991	1973.25	3.13
1992	2264.20	3.22
1993	1667.25	3.32
1994	2057.15	3.50
1995	2556.22	2.54
1996	1799.42	3.64
1997	1451.75	3.18
1998	3301.80	4.30
1999	2375.80	5.06
2000	2433.45	4.01
2001	2178.60	4.39
2002	1917.30	2.40
2003	1943.20	3.36

Source: Anonymous, 2004

Table 14.2 Chi Square Test.

	Value	df	A symp. Sig.(2-sided)
Pearson Chi-Square	156.000a	144	0.233
Likelihood Ratio	66.689	144	1.000
Linear-by-Linear Association	1.852	1	0.174
McNemar Test			
N of Valid Cases	13		

Cropping arrangement is planning alternative opportunities that are sensitive to changes and conflicts that arise. In the months of water deficit, the intensity of conflict increases. Identification of conflict-related decline in water resources will be discussed in the next section.

14.3.2 Water Resource Utilization Conflict

The water resources utilization stakeholders are the local community around Krabyakan Spring, the local community of Pasuruan Regency as the major beneficiary of water resources from Krabyakan Spring (divided into household, industry, agriculture farmer, and embankment farmer), the local community of Sidoarjo Regency as the potential user of water resources from Krabyakan Spring, and also the local government of Malang, Pasuruan and Sidoarjo Regency. Each stakeholder group has its own point of view and interest with regard to utilizing the water resources, and these differences can lead to conflict. The conflict occurred Pls The Malang Government arranged a Memorandum of Understanding (MoU)

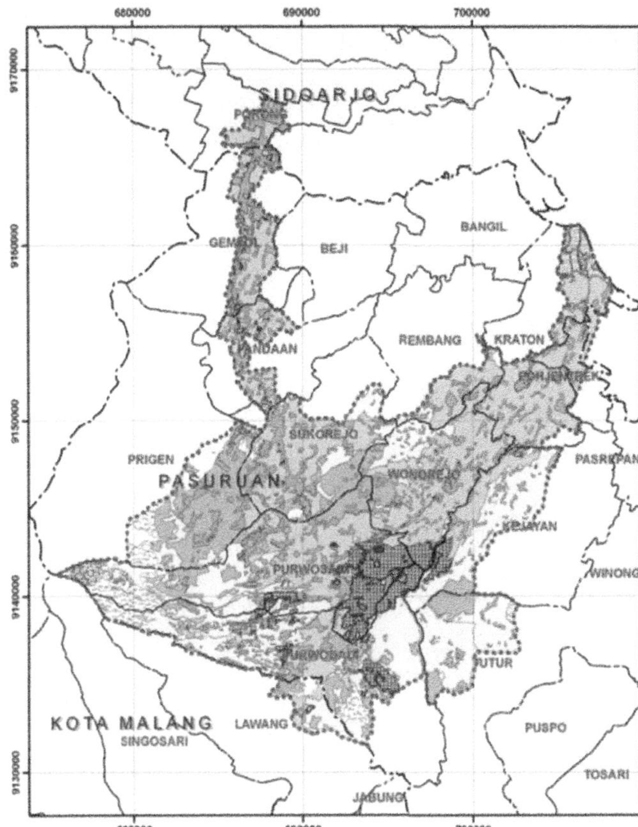

Figure 14.4 Land use of Welang Watershed area.

with Government of Sidoarjo about water resource in this case for utilization of Krabyakan Spring. First of all this MoU were the resource persons triggered by scarcity of water in Sidoarjo because of the Lapindo hot mud flow disaster. In addition there are economic reasons for the increase in revenues of Malang Regency.

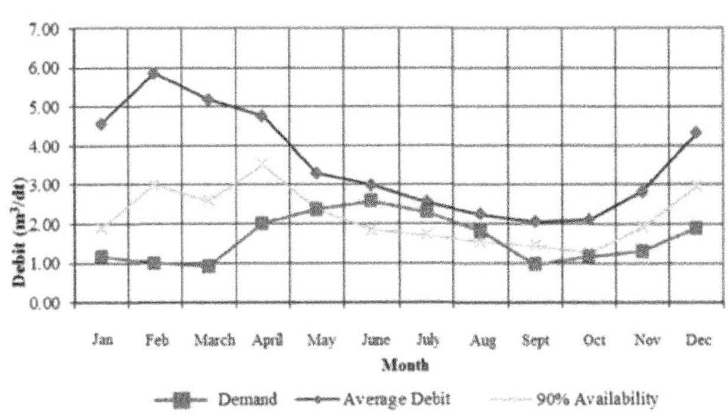

Figure 14.5 Supply demand water balance in Welang watershed.

In the Krabyakan Spring utilization plan, there is a planned pipeline that connects the district of Lawang, Malang regency headed to Porong District, Sidoarjo, and Pasuruan (Figure 14.6). Because of this utilization plan, there were 638.964 people in 16 sub-districts are affected. In the process there is a conflict of interest between stakeholders, especially for people in Pasuruan, which holds the largest-affected communities.

Figure 14.6 Transmission pipe from Krabyakan Spring (Malang Regency) to Sidoarjo Regency.

Conflict mapping is needed to identify the conflict of each stakeholder, which will be composed from the identification of stakeholders, their interest, concerns, conflict form, potency and also weakness related to the conflict that occurred in the study area. This tool is used to identify the condition and as one of the basic considerations to formulate conflict-sensitive strategies related to climate change. An explanation of the conflict mapping can be obtained in Table 14.3.

Based on Table 14.3, conflict of interest among the community is exacerbated mostly because of a lack of coordination among local government regencies, and also low local community participation. In cases of problems related to trans-boundary authority, authority will be delegated to the government at a higher level as a coordinating institution region IV (BAKORWIL Malang, Pasuruan, Jember, Probolinggo) and the government of East Java Province. The role of these institutions is not integrated, so that the conflict remains unresolved.

Table 14.3 Conflict of interest mapping in Krabyakan Spring water utilization.

Stakeholders	Interest & Concern	Conflict	Potency & Weakness
Local Community (Domestic use of water)	The fulfillment of water needs, in sufficient quantity	Conflict with government of Malang Regency	Awareness of community groups is quite high Low opportunity to participate in development
Agriculture Farmer	Demand for water in the dry season is fulfilled so that the farmer achieves harvest success	Conflict with government of Malang Regency	Low opportunity to participate in development
Embankment Farmer	Demand of water in the dry season is fulfilled so that the embankment production sustainable	Conflict with government of Malang Regency	Low opportunity to participate in development
Private stakeholders (industry)	Water needs are met, water availability decreases due to the utilization of Krabyakan Spring	Conflict with government of Malang Regency	Low opportunity to participate in development
Potential water user in Porong, Sidoarjo	Water demand fulfilled	Conflict with government of Pasuruan and local community of Pasuruan	Low opportunity to participate in development
Government of Malang	Provide the public with adequate water supply, and develop economic regions by increasing local revenues	Conflict with local community (agriculture & embankment farmer, industrial and domestic community), and the government of Malang	Have a role as facilitator and enabler, which should connect the interests of society with the development of the utilization of water resources
Local Government of Pasuruan	Provide the public with adequate water supply	Conflict with government of Malang Regency	
Local Government of Sidoarjo	Provide the public with adequate water supply, concern MoU failure of Malang & Sidoarjo	Conflict with government of Pasuruan Regency	Indirectly conflicted with Pasuruan regency, the potency of the conflict is low

14.3.3 Conflict Resolution of Water Resource Utilization and Adaptation to Climate Change Strategies

To address the threat of climate change in relation to the utilization of water resources, and conflicts that occur in the process of utilization, strategies must be formulated. First, there should be a strategy for the technical utilization of Krabyakan Spring water that is tailored to the principle of equity. And the second is that its strategy at the level of decision-makers.

A strategy for the equitable distribution of water to every stakeholder can be executed by regulating the time and capacity based on the equity principle, for citizens in the district of Malang, Pasuruan, and Sidoarjo. Additionally, a model based on supply and demand can be considered an alternative to form a reservoir to hold rainwater that can be utilized in the dry season when there is a deficit in the availability of water. Last but not least, a green belt plan can be formulated to create a conservation area around the springs.

A second strategy is associated with policy. Climate change policies and programming in Krabyakan Spring must be sensitive to conflict dynamics, or there is a risk that response strategies could aggravate tensions and increase the prospect of violent conflict. The facilitator, in this case is the government of Malang and Pasuruan Regency, must try to implement a community-based approach to identify conflict risks and potential solutions, and to provide feedback on the impact of interventions on conflict dynamics. The next strategy is to support and build upon local adaptation mechanisms. The policy that the governments settle on must draw upon traditional adaptation and coping mechanisms. Strategies traditionally employed by the community around Krabyakan Spring to conserve the buffer area of the spring, for instance could provide useful lessons for livelihood security in environments that become increasingly harsh as a result of climate change. Integration of climate change in development policy and strategies is also key to solving conflict related to climate change. Development policies and programming will influence the outcomes of climate change strategies – and the other way around. If climate change policies are coherent with, and reinforce, national-level policies regarding natural resources, land, water and agriculture, they will be more effective in general, and in mitigating the conflict threat in particular. Over the longer term, climate change sensitivity should be mainstreamed into policymaking and programming in all of the identified areas.

As for society, a strategy to improve the participatory approach would be to programs especially for the equitable distribution of water and the conservation of the upper area, using ideas generated by society members, who determine how resources will be shared among themselves and the government. This approach can improve society members' sense of belonging and foster their ability to organize themselves and their resources.

14.4 CONCLUSION

Based on the results of this study, climate change has an impact on water resources availability, significantly correlated, with regression formula y = 2, 2 + 0,001x, and its influence can be seen on water demand and food security and also with the conflict that occurred between several beneficiaries of water resources in the study area. Strategies for conflict resolution regarding water resource utilization and adaptation to climate change formulated from this study are the following: managing water distribution, strengthening the bargaining position of the beneficiaries in society, being sensitive to conflict dynamics in climate change policies and programming in Krabyakan Spring; drawing upon traditional adaptation and coping mechanisms when determining policy; and integrating climate change in development policy and strategies.

14.5 REFERENCES

Anonimous (2003). Malang Regency Regional Development Planning Document *(RTRW)*, Malang Development Planning Agency, Malang, Indonesia.

Smith D. and Vivekananda J. (2007). *A Climate of Conflict: The Links between Climate Change, Peace and War.* International Alert, London.

FAO (2003). World Agriculture : Towards 2015/2030 A FAO Perspective. In. J. Bruinsma (ed.), Earthscan Publications, London.

Houghton J. T., Ding Y., Griggs D. J., Noguer M., van der Linden P. J., Dai X., Maskell K. and Johnson C. A. (eds.) (2001). Climate Change 2001-The Scientific Basis. *Contribution of Working Group I to The Third Assesment Report of the Intergovernmental Panel on Climate Change.* Cambridge University Press, Cambridge, United Kingdom and New York, NY, USA.

Smith D. M. (2003). *Just One Planet: Poverty, Justice and Climate Change.* Practical Action Publishing, UK.

Anonymous (2004). Purwodadi Climate Station Report 1991–2003. Purwodadi Climate Station, Pasuruan, Indonesia.

15 Sustainability of water supply synchronizing PDAM and HIPPAM in Indonesia

Joni Hermana

Department of Environmental Engineering, Faculty of Civil Engineering and Planning, Kampus ITS Sukolilo, Surabaya – 60111, Indonesia. E-mail: hermana@its.ac.id

Abstract

The paper highlights the important question of how the PDAM can be sustained as a water supply operator, since the national drinking water supply coverage is still low, at only 30% in urban and 6% in rural areas in Indonesia. Several problems have been sorted out to maintain the water supply in a sustainable way, including regulations, human resources, debts, raw water quality, water distribution, service quality, corporate culture and community attitudes, particularly with respect to the MDGs by 2015. Taking into account that the government has been unable to fully support the PDAM financially, two approaches are proposed to handle the problems: the downstream approach and the upstream approach.

Keywords: downstream, PDAM, upstream, water supply

15.1 INTRODUCTION

According to ta WHO survey in 2004 (WHO and UNICEF, 2006), 47% of Indonesia's population lived in municipal areas and the remaining 53% livied in rural areas (out of a total population of 220.077 million). By the end of 2004, population with access to improved water stood at 77%, with only 17% of households having a connection to the PDAM (Municipal Water Supply Enterprise). In municipal areas, 87% of the population have access to the water supply, but only 30% of households have a connection to the PDAM. On the contrary, out of 69% of people in rural areas who have access to improved water, only 6% of households have a connection to the PDAM. The access category to the water supply (The World Bank, 2004) indicates that the population with access to improved water comprises 52% using individual resources (such as pumps and wells), 25% using publicly owned resources (water hydrants), and only 15% depending on public or communal utilities (Tumiwa, 2006).

These numbers show that water supply access is still unbalanced, unevenly distributed and potentially insufficient to sustain water supply services in Indonesia. Although the development of access to improved water has almost reached the MDG target, i.e. 52.1% out of 67%, according to the MDG Country Report 2007 (BAPPENAS, 2007), the improved water does not necessarily comply with the quality standard for safe drinking water, most likely because the management of PDAM as the sole water supply agent has not yet been optimized. Until 2004, it was found that out of a total of 318 PDAM organizations, 17% were categorized as healthy, 44% less healthy and 39% not healthy. Data from 2005 showed that only 7% remained healthy, 23% were less healthy, 40% were not healthy and 4% were critical. Problems that commonly affected the performance of PDAM were Unaccounted for Water (UFW) or Non-revenue Water (NRW). These problems were due mainly to high leakage from the distribution piping system, or financial and mismanagement issues. Therefore, there is no leverage to increase the service coverage and OM of the water supply infrastructure.

Millenium Development Goals (MDGs) have stated the aim to reduce by half the population with no access to water and sanitation by 2015 (Target 10: "To halve, by 2015, the proportion of people without sustainable access to safe drinking water and sanitation"), and also the Dublin Conference on Water and Environment has stated that the first fundamental human right is access to water and sanitation with affordable costs. Considering the current condition of our national water supply service coverage, it is estimated that the piping connection for water supply should be increased to serve 92.4 million people, or the equivalent of 40% of the 231 million of our total population. This estimation was based on the initial 39 million, or 18% service coverage, in 2004. The target shows a service coverage increase of 53.4 million people equivalent (PE) within a 5-year period. For the urban population, there will be an increase from 33% (30.8 million) to 66% (61 million), equal to a total of 30.2 million PE with new water supply connections. The increase in water supply to the rural poplulation will begin with a rise from 7% (8.6 million) to 30% (37.7 million), equal to 29 million new water supply connections.

Total water supply production capacity could be estimated based on an increase in new service coverage for 53.4 million PE, the results of which are shown in Table 15.1. Several assumptions were made, such as, water demand is based on our national average water consumption rate, *i.e.* 120 l/cap.day for house connection (SR) and 60 l/cap.day for public tap/hydrant (HU) with the connection SR:HU ratio of 80:20. In accordance with the MDGs, access to water supply will reach 80% (201 million) from 252 million PE in 2015. This coverage consists of 48% with piping systems and 32% without piping systems (NN, 2006).

Table 15.1 Estimation for the development of water production capacity.

No.	User Type	Coverage	Average Water Demand	Water Losses (NRW)	Increased Capacity
		(%)	*(l/cap/d)*	*(%)*	*MLD*
1.	House Connection	80	120	20	6,151,680
2.	Public Tap/Hydrant	20	60	20	768,960
Total Capacity					6,920,640

Source: NN (2006)

The investment needed to develop the increased water supply system is USD 2446.81 million in order to achieve 80% service coverage in urban areas and 40% in rural areas by 2015, or 244.68 million annually (NN, 2006). The total investment was estimated to be around USD 7067.02 million, or USD 542.55 million/year. Nevertheless, the national budget for water supply is only about USD (42.55 – 63.83)/year, far below the required budget.

Taking into account that the government has been unable to support the PDAM sufficiently and that the PDAM is in a condition not conducive to practicing good corporate governance, an important question needs to be raised. Can the PDAM sustain water supply for cities as social capital in Indonesia?

15.2 PROBLEM IDENTIFICATION FOR POTENTIAL INSUSTAINABILITY

15.2.1 Institutional Regulation, Laws and Constitution

Regarding PDAM's water supply system for cities, some laws and regulations at local, regional and central levels overlap, creating contradictions in terms of management, authority or coordination. This overlap tends to weaken the role of the PDAM as the government agent that should serve the community on the one hand and acts as a revenue generator for local governments on the other hand. Further, it was also noted that coordination between local, provincial and central governments is still fragile and in a fluctuating condition.

PDAM management should be allowed to propose local regulations for its own development purposes in order to accommodate its interests in building up systems in more flexible atmospheres, including eliminating the bureaucratic procedure that all decisions must be made with the acknowledgement of the DPRD (local people's representative assembly). PDAM organizations should be granted a more autonomic role in regulating their own internal management systems, such as in determining the composition of their Board of Directors, their relations with the Supervisory Board, and their staff and employees.

15.2.2 Human Resources Empowerment

Human resources are a problem for PDAM management, because staff members are appointed and not all are professionals. Some are government officers or civil servants assigned to work for PDAM. This condition contributes to the low performance of PDAM since such staff members do not always share the enterprise's commitment to serve the company and maintain stewardship of its assets.

There should be a systematic and strategic approach to managing human resources, particularly for 'half-status employees,' such as the appointed staff mentioned above and underclass labourers who have been hired for several years and subsequently demand permanent status. A recruitment system that has lasted without transparency and accountability for years creates burdens for PDAM management, because the number of incompetent staff members and unskilled labourers has reached such significant numbers that, due to social and political risks, the problem cannot be easily sorted out.

Therefore, PDAM management, which has been given these unfavourable conditions, should be able to adjust balances between PDAM's role and the competency of human resources in every sector to optimize performance.

15.2.3 Burden of Debts

Inherited debts from the previous management and interest on these debts are the main concerns in the survival stage of PDAM. This is because less than 9% of PDAMs are classified as healthy in Indonesia. It is therefore almost impossible to develop most of the PDAMs without the support of government funding, although the government has a very limited budget. Increasing tariffs would not be a good strategy. Not only are tariff increases a sensitive matter, but also there is not enough proof of quality service performance to consider such increases.

According to Tumiwa (2006), the government tried to sort out these problems by introducing a Policy Matrix for PDAM Recovery. The policy has 2 main programs, namely program aid and technical aid, with 4 different action plans, namely:

a. Corporation for 96 PDAMs
b. Technical for 74 PDAMs
c. Water Tariffs for 68 PDAMs
d. Costing/Budgeting for 109 PDAMs

Further, the World Bank (2004) made several recommendations for the water sector in Indonesia, such as reformation schemes for PDAM, including debt restructuring, PDAM corporations, the possibility of Private Sector Participation (PSP) identification, and PDAM mergers. The World Bank also proposed developing budgeting of work frames (including for the private sector), identifying sources and making budget strategies for the cities.

Considering their own limited budget, the government tried to ameliorate the PDAM's difficult situation by promoting the KPS (Government Private Cooperation) or PPP (Public Private Partnership) in order to increase water supply coverage and achieve the MDGs. However, for almost two decades, private sector participation made no significant positive contribution to achieving the targets, and showed no superior differentiation from the public sector (Tumiwa, 2006).

15.2.4 Raw Water Quality and Distribution Water

The quality of raw water becomes a thread to the PDAM itself. Some variables affect its reliability, including raw water resource management, environmental protection and

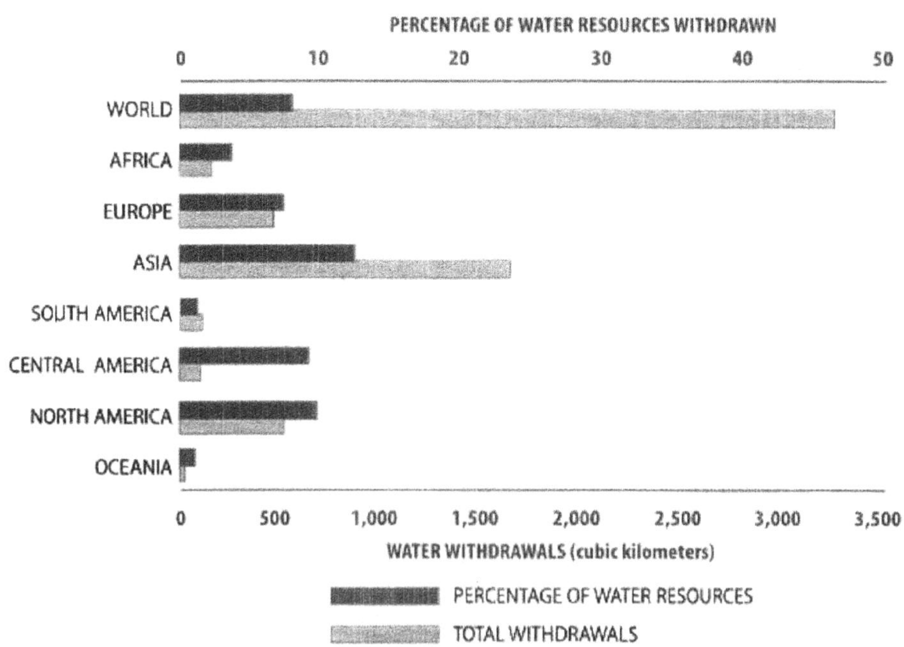

Figure 15.1 Increasing water demand with decreasing supply from 1987 to 1995 (WRI, 1998).

maintenance responsibility, and intersectoral coordination. The quality of distribution water depends greatly on the applied treatment processes and technology, operational and maintenance of distribution pipe networks in the city.

Water demand increases proportionally to population growth; conversely, water resources decrease continuously (Figure 15.1).

In Indonesia, the availability of water resources is in fact quite abundant, with an average flowrate of 15,165 m^3/capita/year, while for individual islands it varies from 1,582 m^3/capita for Java and Bali to 418,803 m^3/capita for Irian Jaya (Ramu, 2004). The demand in 2000 for irrigation and Domestic-Municipal-Industry (DMI) for Java/Bali was about 480 m^3/capita (equal to 1,777 m^3/s), which is about 30% of the surface water potential (Table 15.2). Unfortunately, there is an imbalance between availability and distribution, which is highly influenced by the seasons. Particularly, in the most densely populated islands, such as Java and Bali, water demand far exceeds supply, and water is also heavily polluted. Java island has only 4.5% of national water resources, but 59% of Indonesia's population inhabit the island. Such a burden results in a water crisis, especially during the dry season, where the river is estimated in a low flow (Table 15.2). The average rainfall is over 2,500 mm/year, of which 80% falls during the rainy season (October to April). The annual renewable water resources are estimated to be about 3,805 km^3 (97,825 m^3/s), while the total estimated fresh water demand in 2000 was only about 2,732 m^3/s (with 90.8% used to meet irrigation demands).

Table 15.2 Surface water resource potential and demand by island, Indonesia.

Island	Area (1000 Km²)	Potential Water Surface (m³/s)	Estimated Low Flow (m³/s)	Irrigation + DMI Demand 1990 (m³/s)	Irrigation + DMI Demand 2000 (m³/s)	Irrigation + DMI Demand 2015 (m³/s)	Yr. 2015 Water Resource Utilization %
Jawa/Bali	139	6,199	786	1,074	1,777	1878	29.8
Sulawesi	187	2,488	561	126	365	529	21.3
Sumatera	470	23,660	4,704	297	497	693	2.9
Kalimantan	535	32,279	6,956	73	93	193	0.6

Note: Irrigation demand is in the range of 87% to 95% of the total demand.
Source: After Ramu (2004)

On the other hand, water demand increases annually and is estimated by the Government to reach approximately 220% by 2020 (NN, 2008). The estimated 2020 demand for irrigation, DMI, fishponds, livestock, and river maintenance was then compared with the estimated natural basin discharge. It was estimated that for Java and Bali (56% of the population) nearly 60% of the natural basin discharge will be required in 2020 to meet the demand, while for Kalimantan (1.8% of the population) the figure is only 1% of the natural basin discharge (UNDP/FAO study in Ramu, 2004).

Taking into consideration the quality of raw water, the main problem of drinking water production by PDAMs is very common at the national level, *i.e.* raw water resources are polluted by both domestic and industrial wastewater. For example, the Surabaya River, the main raw water resource serving more than 2.3 millions people of the Surabaya Municipality, is a branch of the Brantas River, and it crosses from

Lengkong Dam in Mojokerto to Surabaya City, as far as 42 km away (Hermana and Razif, 2004). Within the vicinity of the water catchment area, there are various scale industries, about 42 of which are under periodic monitoring because of their potential pollution into the river. Besides, the most densely populated area is also located along this catchment area. These conditions contribute significantly to pollution problems in the Surabaya River, where almost all pollution parameters exceed the quality standard, such as BOD concentration (Figure 15.2).

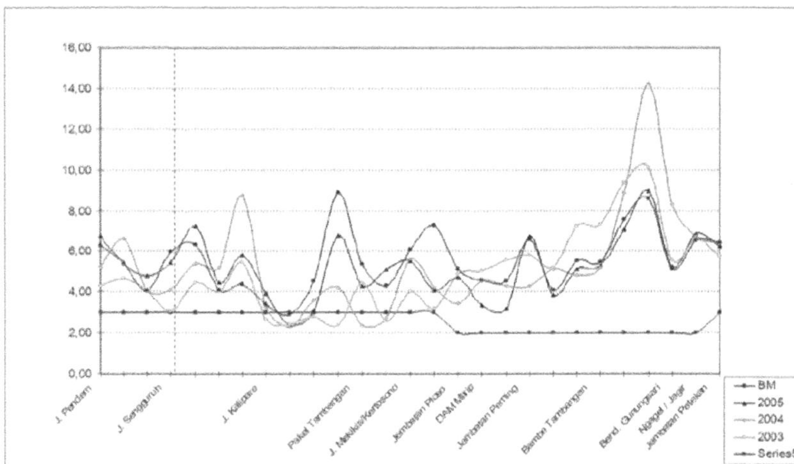

Figure 15.2 The BOD Concentration along the Surabaya River compared to the standard (BM).

The degradation of raw water quality reduces community trust. Conventional treatment processes are considered inappropriate for removal of the most recent pollutants. As a consequence, since the beginning of the 21st century, world bottled water consumption has increased significantly, at an annual rate of 7% (Figure 15.3; Belot, 2000 in Verstraete, 2004). In 2004, Indonesia was the 8th country in the world in bottled water consumption, with an average of 7.62 million m^3/year, although bottled water is almost 200 times more expensive than PDAM water. Bottled water consumption could potentially erode the capability of the community to achieve self-prosperity (Wignyosukarto, 2008); instead, the community enriches foreign investors, because almost 65% of the Indonesian bottled water market is owned by international companies, especially Danone and CocaCola (Tumiwa, 2006)

On the other hand, there are some pollutants that cannot be treated by conventional water treatment methods and could potentially harm human health (Verstraete, 2004), such as the following:

a. Unwanted organism propagulates (examples: *Cryprtosporidium, Legionella)*
b. Trace chemicals (examples: THM, nonylphenol, EE2, atrazine, pharmaceutical products)
c. Excess nutrients and pesticides from extensive agriculture
d. Heavy metals, such as *lead* (Pb), *arsenic* (As), *chromium* (Cr), and possibly *mercury* (Hg).

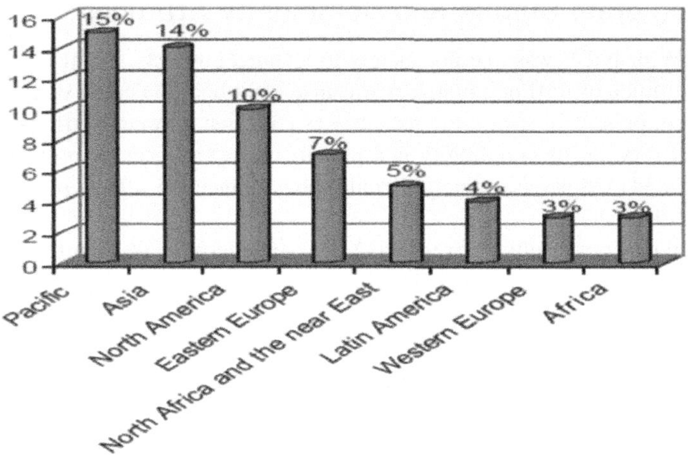

Figure 15.3 The growth of bottled water consumption in the world 1999–2001 (Belot, 2000 in Verstraete, 2004).

Therefore, to treat water safely, it is necessary to apply sophisticated technology, such as membrane technology, according to the spectrum of pollutants in the raw water (Figure 15.4).

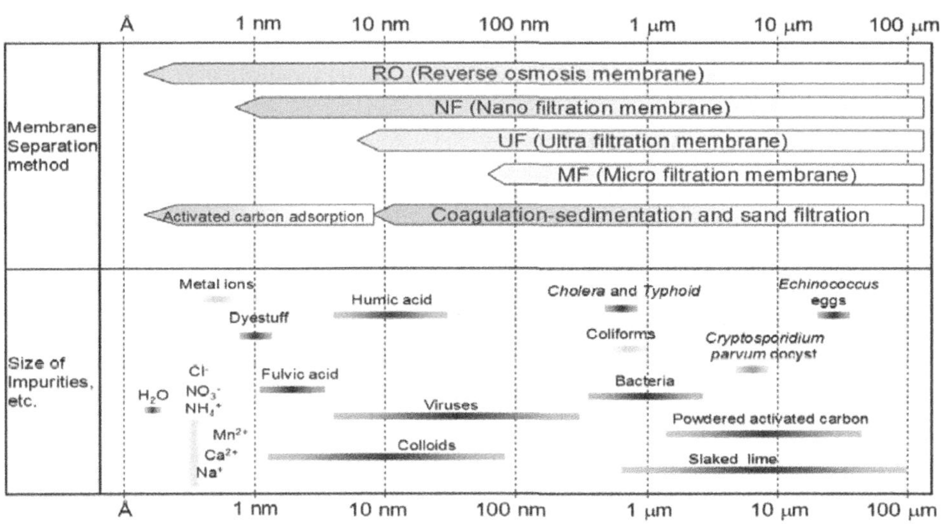

Figure 15.4 Ability of membrane as alternative technology for removing different impurities from raw water.

15.2.5 Service Quality

PDAM's service quality has not met expectations. There are at least two problems: (1) the uncertain handling of demand for new house connections, and the handling of complaints; and (2) the response time of PDAM management when dealing with problems related to service quality.

15.2.6 Corporate Culture and Community Attitudes

The high UFW or NRW was due not merely to technical aspects, but also to nontechnical aspects. The attitude of staff or employees was also identified as a cause when high UFW or NRW occurred, present in such circumstances as collusion between staff and customer, or ignorance that serving the community is a priority. As a result, many problems occur that could endanger PDAM assets, namely illegal piping connection, illegal pumping, and even physical asset destruction.

An internal corporate culture for the PDAM as an enterprise was hardly recognized; the PDAM was just quite bureaucratic. There was no spirit of competition, entrepreneurship, awareness or empathy within the company.

15.3 APPROACH TO SUSTAIN THE WATER SUPPLY FOR THE CITIES

The water supply problem could be addressed with two differing approaches: the *downstream approach* and the *upstream approach*. The main problem with downstream is the imbalance between supply and demand, raw water pollution, and water resources management. The upstream problem was generated from un-integrated, even overlapping, institutions and regulations, the lack of an internal management system, and issues with human resources. Both approaches are complicated because of little financial and investment support from the government to the PDAM.

Several scenarios were examined to overcome the problem either upstream or downstream in order to sustain the capability of the PDAM to provide water supply for the cities of Indonesia.

15.3.1 Upstream Approach

1. There should be more efforts taken by the government to enhance economic growth for the whole country so that population density will be more balanced, resulting in more reliable distribution of the water supply.
2. There is an urgency for the government to readdress and revaluate the concept of regional planning, taking into account the RTH (green open space) ratio, water reserves using phytotechnology and ecological concepts, developing environmentally friendly schemes through mitigation and adaptation mechanisms in order to be more active in climate change progress by means of the CDM concept.
3. There should be more mutual cooperation between provincial and local governments to regulate the assimilative capacity and carrying capacity for all water bodies or resources. This is to assure water resource availability for a long-term period.

15.3.2 Downstream Approach

1. Each government at different levels should establish institutions that work integratedly to determine the quality and quantity of raw water resources for drinking water, including setting standards and mapping raw water demand.
2. It is necessary to improve the professionalism of human resources and management of the PDAM by involving national private companies and the community to participate in the water supply sector. To some extent, the KPS (Government Private Cooperation) or PPP

(Public Private Partnership) could be applied with some corrections since it was reported that for almost two decades, private sector participation showed no significant positive contribution toward achieving the targets, nor did it differentiate itself as superior to the public sector (Tumiwa, 2006).

3 Technology options, well-trained operators and community participation play important roles in the water supply system. This is mainly because financial support from the government is insufficient. To meet targets, enhancing community-based water supply programs, such as HIPPAM, could be a wise strategy under current conditions. It should be noted that although access to water at affordable costs is the first fundamental human right according to the Dublin Conference, it must also be considered that all community components, proportional to their roles and capacities, must take responsibility for maintaining the sustainability of the water supply system and distribution. The head of the local government should be responsible for the existence and maintenance of water resources to assure quality control, as well as ecological and cultural functions in regions over which they have authority.

4 It has not been an economic option for some PDAMs to handle their pollution problems by trying to relocate their raw water dams, because such relocation would not affect the raw water quality extensively. The more economic solution would be a relocation of intake pipes or even treatment plants (WTP), in addition to continual efforts to control main pollution sources from industrial, domestic and agricultural activities along the water bodies.

5 In order to overcome financial constraints, there should be a technology breakthrough for water treatment that is affordable from an economic point of view, to control industrial and domestic wastewaters along the rivers, especially for communal treatment scales. Domestic wastewater is becoming a major concern due to its pollution potential to water resources, both groundwater and surface water.

6 Finally, revitalization of regulator and executor bodies, such as BPSPAM, Jasa Tirta and Dinas Pengairan would be an advantage for raw water resources management that could help the PDAMs. This integrative approach is to manage the river basin development plan that regulates water usage for the water supply and allocates wastewater disposal loads that must comply with standards, based on assimilative and carrying capacities.

15.4 REFERENCES

BAPPENAS (2007). Report on the achievement of millennium development goals Indonesia 2007. National Development Planning Agency, Republic of Indonesia.

Hermana J. and Razif M. (2004). Reevaluation on the suitability of Kali Surabaya as raw drinking water of Surabaya municipality in Indonesia. *Proc. of Conference on Water & Wastewater, Kuala Lumpur.*

N. N. (2006). *Pengembangan sistim penyedian air minum* (The development of drinking water suply system). Direktorat Pengembangan Air Minum Dit. Jen. Cipta Karya. Public Works Department. Indonesia.

N. N. (2008). *Pemerintah akan meningkatkan penyediaan air baku* (The government will increase raw water supply), https://www.pu.go.id. [Accessed: 2009, March 18].

Ramu K. V. (2004). *Brantas River basin case study Indonesia: Background paper. agriculture and rural development*, Word Bank.

The World Bank (2004). Averting an infrastructure crisis: a framework for policy and action. The World Bank East Asia – Asia Infrastructure Department and Indonesia Country Program.

Tumiwa F. (2006). Alternatif sumber dan mekanisme pembiayaan infrastruktur air di Indonesia, (An alternative resources and mechanism of water infrastructure financing in Indonesia), Symposium "Securing the Right to Water in Indonesia", Indonesia.

Verstraete W. (2004). Environmental Biotechnology: The European Perspective. International Seminar on Up-to-date Water Treatment and Reuse, The Center for Environmental Sanitation, University of Ghent, Belgium.

WHO and UNICEF (2006). Meeting the MDG drinking water and sanitation target: the urban and rural challenge of the decade. WHO/UNICEF, Joint Monitoring Programme for Water Supply and Sanitation.

Wignyosukarto B. (2008). Pengelolaan Sumberdaya Air Terpadu dalam Upaya Pencapaian Tujuan Pembangunan Milenium 2015 (Integrated water resources development to reach Millennium Development Goals (MDGs) (2015). In: http://budiws.wordpress.com/2008/01/09/ [Accessed: 2009, 28 January].

WRI (1998). A Guide to the Global Environment. World Resources 1998–99.

WRI, UNEP, UNDP, World Bank (1998). *World resources 1998-99: a guide to the global environment - environmental change and human health*. Washington, D.C., The World Bank. http://documents.worldbank.org/curated/en/1998/01/442702/world-resources-1998-99-guide-global-environment-environmental-change-human-health.

Index

IWA Publishing's authorised EU representative for General Product Safety Regulations is Diane D'Arras, 15 rue Duret, 75116 Paris, France, e-mail: safety@iwap.co.uk.

Printed and bound by CPI Group (UK) Ltd, Croydon, CR0 4YY
12/05/2026
02108891-0009